Perceptual-Motor Development Equipment

Perceptual-Motor Development Equipment

INEXPENSIVE IDEAS AND ACTIVITIES

PETER WERNER
University of South Carolina
Columbia, South Carolina

LISA RINI
Washington State University
Pullman, Washington

JOHN WILEY & SONS, INC.
New York / London / Sydney / Toronto

Library of Congress Cataloging in Publication Data:

Werner, Peter H
Perceptual motor development equipment.

Bibliography: p.
Includes index.
1. Perceptual-motor learning. 2. Teaching–Aids and devices. I. Rini, Lisa, joint author. II. Title. [DNLM: 1. Child development–Motor skills. 2. In infancy and childhood. 3. Perception–In infancy and childhood. 4. Equipment and supplies. WS105 W494p]
LB1067.W47 '372.1'3' 75-43744
ISBN 0-471-93371-6

Printed in the United States of America

10 9 8 7 6 5 4 3 2 1

Preface

Over the last decade perceptual-motor learning programs have run the gamut in terms of acclaim toward the development of physical, intellectual, emotional, and social capacities of children. Some researchers have suggested that perceptual-motor activities constitute panaceas for children with mild and severe learning disabilities. Others have refuted the point by stating that special perceptual-motor programs have little or no effect on the development of such children. Much more research needs to be conducted before any definite conclusions are reached. However, it has been established that perceptual-motor activities constitute an important part of the experiences that help *all* children prepare for school. There are certain low-level foundational or readiness skills involving perceptual-motor learning that, when competently handled, will prepare *all* children for future school tasks, such as cognition, reading, writing, and so on.

Most information, books and articles, dealing with perceptual-motor development, concerns the theories of child development and success-oriented activities for children. Specialized equipment has been designed by perceptual-motor specialists and commercial manufacturers for use in these programs. However, teachers in the public schools and in preschool settings often find inadequate conditions. Teachers, confronted by a lack of funds with which to purchase equipment for children's use to aid in their physical, intellectual, emotional, and social development, become frustrated and lose their drive or motivation to provide children with the best experiences possible to enhance their development.

Having had the experience of operating under these conditions as the directors of a perceptual-motor laboratory, the writers set about a plan of constructing homemade, inexpensive equipment to meet children's developmental needs. Various equipment ideas began to multiply. They were shared with students, teachers, and parents who also became motivated to begin constructing their own equipment. As a result, the stimulation needed to write this book, evolved.

To The Teacher-Parent

For those of you who are working with young children as teachers, tutors, or parents, this book is intended to be a source of ideas for equipment and activities that will help children develop foundational perceptual-motor skills. The beginning of each chapter is a brief description of the concepts related to the topic. Following the introduction to each chapter are selected pieces of equipment for use in perceptual-motor development that can be made by you or your children. Ideas for constructing equipment, as well as objectives and suggested activities for each piece are also included.

To The Student

For those of you who are students in learning disabilities, perceptual-motor development, special education, adapted physical education, physical therapy, or occupational therapy, this book is intended to be a supplementary text for the unification of theoretical foundations and practical application in a developmental program. As you learn about the various theoretical approaches and concepts relating to learning disabilities, use this book as a reference to learn about the equipment and activities that can be used to help children develop specific perceptual-motor concepts.

Special thanks go to the students working in the Perceptual-Motor Laboratory at Miami University, Oxford, Ohio, fellow teachers, and parents who have contributed ideas for equipment construction and use. The writers would also like to thank Jane Holloway and Mary Werner for creating the artwork, and finally, Andrew Ford, Jr., Wayne Anderson, Rochelle Sherwood, J. Frances Tindall, Erica Nartasi, Penny Doskow, and the many people at John Wiley Publishing Company, Inc., who have contributed to the completion of this project.

Peter Werner
Lisa Rini

COLUMBIA, SOUTH CAROLINA
PULLMAN, WASHINGTON
SEPTEMBER 1975

Contents

1

Introduction

WHAT IS PERCEPTUAL-MOTOR DEVELOPMENT?

WITHIN RECENT years educators, researchers, and clinicians including parents have taken an interest in the use of perceptual-motor activities in education programs for preschool and elementary children who are classified as "normal" as well as those who are handicapped in any way. Although volumes have been written on the topic, research still remains somewhat inconclusive. There are some generalizations, however, that permeate the literature. They will be briefly discussed here for purposes of setting the atmosphere for the contents of this book.

The definition of perceptual-motor is inherent in the learning process (6, 22, 27, 34, 41). At first, there is a stimulus or stimuli to our senses (visual, auditory, tactile-kinesthetic, smell, and taste), which is called receiving or decoding the message (Figure 1.1). After the initial stimulus, the message is carried through the afferent nerves toward the brain. This is called input. Once the message reaches the brain or cortical level, a process of organization and integration occurs.

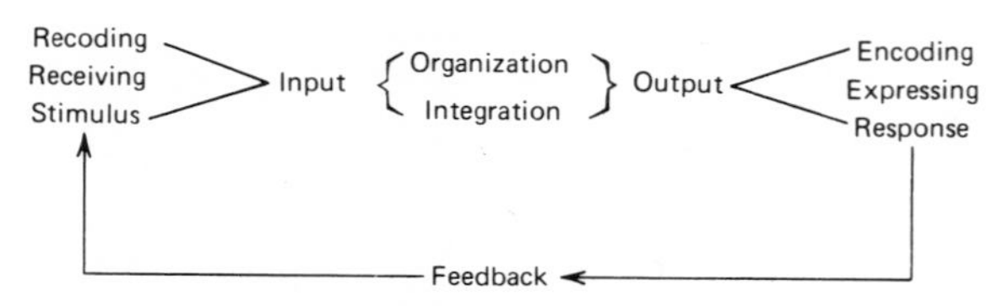

FIGURE 1.1. The perceptual process.

Current messages are interpreted by combining present information with past experiences and decisions to act are made. The decisions are carried, via the efferent nerves, toward one of our response mechanisms. This is called output. Finally, the response is made in the form of vocal expression or a motor act (i.e., running or writing). After the response is made and we observe the results, a feedback mechanism filters off this knowledge and feeds it back to the stimulus portion of the cycle, causing the generation of modified stimuli and inputs. If the response is correct and successful, the brain retains this information for later use. If the response is incorrect, an altered response will be made until the response is correct or until the individual surrenders to his frustration or failure. Thus the acts of perception, followed by a motor act, are mutually dependent. It is extremely important that a child develops an efficient perceptual-motor processing system for successful task performance.

STAGES OF DEVELOPMENT

Research and theories concerning child development indicate that children pass through various stages of growth as they seek maturity (1, 2, 27, 39). During infancy and early childhood children pass through the sensory motor or gross motor stage, during which they experience and learn about their environment through the motor domain (reach, grasp, release, balance, crawl, creep, walk, taste, etc.). (Figure 1.2) Gradually, the child progresses to a perceptual stage where his or her primary means of learning is usually through visual perception. The auditory and tactile-kinesthetic senses, however, also play an important part in this stage. Both of these stages lay the foundation or create an ability to achieve at the cognitive level, the concepts and skills related to symbolization, abstraction, verbalization, reading, and other academic tasks. Thus, it is important for children to receive a solid background of motor and perceptual experiences to serve as a foundation for school learning.

Because of the success-failure ratio and stages of development, it has been demonstrated that perceptual-motor development programs contribute to the child's body image or self-concept (11, 12, 27). A program geared to the individual's functioning level will help him achieve success. As the child succeeds, he will be challenged and motivated to attempt other tasks. He will gain self-confidence and will try more difficult tasks. Thus, he will ascend toward the cognitive learning level. A child who consistently fails will stop responding and will experience difficulties in learning.

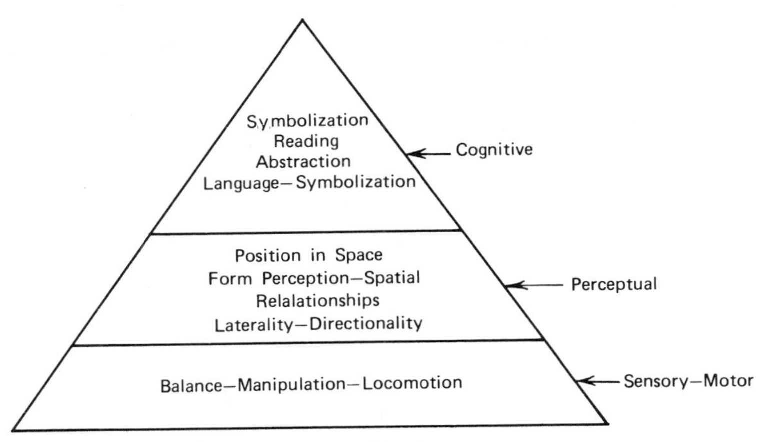

FIGURE 1.2. Stages of development.

DIAGNOSIS

In order to initiate a perceptual-motor program, the students' problems must be detected. To serve this purpose, there are many tests available for screening children at various stages of development. For example, the Purdue Perceptual-Motor Survey and Ayres' Southern California Perceptual-Motor Tests are designed to screen children at the sensory-motor stage of development. The Frostig Developmental Test of Visual Perception and Wepman Auditory Discrimination Test are intended to screen children at the perceptual stage of development. The Illinois Test of Psycholinguistic Abilities screens children at the cognitive stage of development. These tests should be administered to determine the child's performance level and to detect individual problems. However, because of a lack of familarity with the tests, lack of time, or insufficient funds to purchase the tests, teachers often fail to do so. Therefore, the authors have included a brief device to be used for initial screening. All the tasks, except the eye-pursuit tests, are common activities. They are easy to administer in a relatively short period of time, and because they are free, they are consistently available to all teachers. Therefore, the examiner should be able to distinguish among levels of performance quality. For best results, children should be screened individually.

This screening test is a means of exposing perceptual-motor problems. It is not a measure of specific performance rated on a scale. However, if a child experiences difficulty in performing these tasks, the teacher should administer one or more of the specific standardized tests to cite the problem areas and place the child in a perceptual-motor program designed to meet his individual needs.

The reader should study other sources for an in-depth knowledge of these diagnostic tests and of the theories of learning disabilities and perceptual-motor learning. The main emphasis in this book is on equipment construction and use in perceptual-motor programs.

SCREENING TEST

Item	Observation	Rating		
		Yes	No	Comments
Body Parts	Can the child identify each body part: eyes, feet, shoulders, wrists, knees, elbows, ankles, chest, hips, and back.			
Touch One Body Part To Another	Touch right hand to left ear. Touch left hand to right foot. Ask two or three others. Can the child show you correctly each time?			
Balance On One Foot	Can the child balance on one foot with his eyes open for about 10 seconds and with his eyes closed for about 3 seconds?			
Hopping	Can the child hop in one place without losing his balance ten times on each foot?			
Jumping	Can the child jump up with both feet at the same time? Can the child jump out with both feet as in the long jump?			
Skipping	Can the child skip leading with alternating feet and arms?			
Balance Beam	Walk forward, backward and sideways across the beam. Does the child control his balance at all times?			
Jumping Jacks or Angels in Snow	Can the child move his arms together? Feet together? Vary the pattern you ask the child to perform.			

Item	Observation	Rating		
		Yes	No	Comments
Tie Shoes Put on Clothes	Does the child have unusual difficulty tying his shoes or putting on his clothes?			
Draw Between The Lines	Can the child hold his pencil and draw neatly between the lines on paper and pencil tasks?			
Draw Forms	Can the child reproduce an accurate copy of the following forms when they are shown to him? Circle, cross, square, triangle, divided rectangle, diamond.			
Block Pattern Visual Sequence	Design a pattern of blocks. Can the child reproduce the same pattern of blocks by using the same shapes and by placing them in the same sequence?			
List of Words Auditory Sequence	Say a list of three or four words to a child and ask him to repeat them. Use common words and do several sequences. Can the child repeat the sequences correctly?			
Auditory Discrimination	Say words that sound similar to a child and ask him to tell you when they are the same or different. Bean-mean, said-lead, real-feel, game-game, and so on.			
Attention Span	Can the child attend to a task for a reasonable length of time? Is he easily distracted?			

Item	Observation	Rating		
		Yes	No	Comments
Ocular Pursuit	Can a child follow a moving target in a horizontal, vertical, diagonal, and rotating direction? The head should be kept still while only the eyes move. Do both eyes; left eye alone and right eye alone.			

Thus, children should be diagnosed properly and helped in each one's specific, weak areas. After a child is diagnosed, it is essential that an individualized program be formed to best meet his needs. Remember: (1) to find and begin at the child's functional level and work toward the complex; (2) to progress from the motor and perceptual levels toward the cognitive; (3) to find and use the child's strongest sensory modes (auditory, visual, tactile, and kinesthetic) in order to build a pattern of success and achievement; and (4) to gradually work on the child's weaknesses until he achieves a normal developmental level.

DEFINITION OF TERMS

While there is no direct evidence that perceptual-motor activities improve intellectual performance, there are certain components of the educational program to which physical activities can contribute in a constructive way (15, 37). These items are referred to as readiness skills, foundational levels, or, in the case of children with learning disabilities, low-level functional deficits. Some of them are listed and defined as follows to help the reader identify specific concepts that children need help with in a developmental context.

Accommodation—the process by which the eye brings an object into focus.

Agraphia—a cerebral disorder marked by an unusual difficulty in learning to write (may be partial or total inability).

Alexia—inability to understand the written language; cannot read with the speed and skills that are characteristic of a particular mental or chronological age.

Aphasia—a delay in speech development that is not a result of deafness or of a defect in the peripheral speech mechanism.

Apraxia—an unusual difficulty in learning patterns of skilled movement sometimes involving physical activity as well as manual dexterity.

Association—the ability to relate visual and or verbal messages in a meaningful way.

Balance—the ability to maintain or regain one's posture or equilibrium.

Bilaterality—using both sides of the body in simultaneous or parallel movement.

Binocularity—the fact that our eyes, though they are set apart in our head, must function together. Each must retain its position with a complex set of covariant postures needing to be established and integrated depending on the object on which the eyes are trying to focus.

Body Image or Body Awareness—knowing the names, movements, functions, and location of body parts; sensing that one side of the body is different from the other sides. It includes impressions received from internal and social feedback.

Closure—completing a behavioral or mental act; the tendency to stabilize or to complete a situation. Closure may occur in any sensory modality.

Convergence-Reconvergence—the occular-pointing mechanism by which the eyes focus upon a target. It enables one to see a single object at varying distances.

Crawling—moving in a prone position. This is done in homologous, homolateral, or cross-pattern movements.

Creeping—moving on the hands and knees. This movement may be done in homolateral or cross-pattern movements.

Cross-pattern—moving the opposite arm and leg at the same time.

Decoding—the ability to understand and comprehend spoken and written words, and pictures: the reception of auditory, visual, tactile, and kinesthetic messages.

Depth Perception—the aspect of visual perception that includes direct awareness of the distance between an object and its' observer: perceiving distance between the front and back of an object, thereby depicting it as three dimensional: the ability to perceive the third dimension in a two-dimensional picture.

Directionality—an awareness of laterality and a projection of this awareness into space.

Dominance—using one eye, hand, foot, or side of the body in favor of the other (may also refer to the different sides of the brain).

Dynamic balance—maintaining equilibrium while performing a locomotor task.

Dyslexia—partial inability to read or to understand what one reads, either silently or aloud. This condition is usually associated with brain impairment.

Egocentric Localization—the ability to locate the position of an object in relation to yourself.

Encoding—the ability to express your ideas through speech or posture.

Eye-Hand Coordination—the ability to integrate sight perception efficiently with a motor act.

Earthbound—an inability to hop or jump; usually due to a lack of balance and/or coordination.

Figure-Ground—the ability to select the primary stimulus from a background of other stimuli. This relationship may occur in any sense modality.

Fine-Motor—the ability to control the small muscles of the body, primarily the eyes and hands, necessary to complete academic tasks.

Fixation—the eyes' ability to maintain focus upon an object.

Form Perception—the ability to perceive an arrangement or pattern of elements or parts constituting a unitary whole. Understanding the relations of the parts to that whole.

Gender Identification—the process whereby a child identifies with the appropriate sex. Research has shown that children who have trouble identifying with their sex, are disabled in learning situations.

Gross-Motor—the ability to use the large body muscles (arms and legs) in smooth coordinated movement for task completion.

Homolateral—moving the limbs on one side of the body in unison.

Homologous—Moving the arms or legs in unison.

Hyperactivity—excessive activity or output in which the child has a surplus of energy and is unable to control movements.

Hyperdistractability—a child's inability to block out incoming sensory stimuli in an effort to perceive the primary stimulus.

Hypoactivity—pronounced absense of activity.

Kinesthetic—an inner awareness of the location of the body parts and how to move them by recalling previous experiences.

Laterality—an inner sense that one side of the body is different from the other. (*See Body Image*)

Learning Disability—children with learning disabilities are those (1) who have educationally significant discrepancies among their sensory-motor, perceptual, cognitive, academic, or related developmental levels that interfere with the performance of academic tasks; (2) who may or may not show demonstrable deviation in the central nervous system functioning; and (3) those disabilities are not secondary to general mental retardation, sensory deprivation, or serious mental or emotional disturbances.

Listening Skills—combination of the ability to hear, an interpretation of sounds heard, and a response to those sounds.

Locomotor—the means of moving from one place to another.

Midline—the movement of the eyes, a hand and forearm, or a foot and leg across the body's midsection without involving any other body part.

Occular Pursuits—the eyes' ability to follow a moving target.

Perceptual-Motor Match—the process that integrates the clues provided by the senses with the responses of the neuro-muscular system.

Perseveration—the inability to develop a new response to a new or altered stimulus. Continuing to behave or respond in a certain way after it is no longer appropriate.

Position In Space—the direct awareness of the spatial properties of an object, especially in relation to the observer. The perception of position, direction, size, form, and/or distance by any of the senses.

Perceptual Constancy—the tendency for a perceived object of a given size, weight, shape, form, and so on to remain constant irrespective of the picture that the retinal image receives.

Receipt-Propulsion—the activities associated with a child's making contact with an object moving towards him and, in turn, the movement that is imparted to an object (catching and throwing).

Sequential Memory—the ability to reproduce a set of visual, auditory, or tactile stimuli from memory.

Spatial Awareness—an understanding of one's relative position in space and the relationship of spatial objects to one another. The ability to see similarities in shape, size, and so on, of two or more objects.

Space-Time Relationship—the ability to translate a simultaneous relationship in space to a serial relationship in time or vice-versa.

Splinter Skill—a skill developed to solve a specific motor problem. It has little or no carry-over value to other motor activities.

Static Balance—maintaining equilibrium while performing a nonlocomotor task.

Visual-Motor Ability—the ability to visualize and to assemble material from life into meaningful wholes; the ability to see and to perform with dexterity and coordination; the ability to control body or hand movements in coordination with visual perception.

Visual Pursuit—following a moving object, keeping it in the center of the field of vision, and providing a direct correlation between the object's movement and the perceptual alterations.

The writers would like to emphasize that it is very important for the readers to gain a knowledge of child development, the stages of motor-perceptual-cognitive development, and diagnostic techniques before initiating a program of perceptual-motor learning. Only when one understands the concepts identified above can the child be best aided in his development. It would be a mistake to use this book haphazardly as a source for equipment ideas without the knowledge of these concepts as background information.

USES OF EQUIPMENT

There are many home items that can be used as developmental perceptual-motor equipment. Teachers should encourage children to bring these various items from their homes. Collection boxes should be provided to keep them assorted according to use. Other "throw away" items, such as appliance boxes, tires, inner tubes, and lumber scraps are often donated by local department stores, service stations and lumber companies. Supplies needed for the construction of low-cost equipment can often be purchased as a school discount from local businesses.

Construction of these pieces of equipment can be implemented in several ways. Some of them can be made by the children or in conjunction with the special education, art, and physical education teachers. Other items, such as geo-boards, sand-paper letters, felt boards, and puzzles, may be made by concerned parents or volunteer groups from the PTA (Parent-Teachers' Association) or ACLD (Association for Children with Learning Disabilities). Students in home economics sewing classes can make equipment, such as letter and number beanbags or manikins for body image projects. Industrial arts classes in shop and woodworking may construct items, such as balance boards, balance beams, rocker boards, or marble tracks. All that is needed is time, effort, concern, and commitment from the children, parents, teachers, and the community!

The remaining chapters of this book are concerned with the development of foundational perceptual-motor concepts, specifically relating to Balance, Body Image, Spatial Awareness, Form Perception, Visual Perception, Auditory Perception, Tactile-Kinesthetic Perception, and Eye-Hand, Eye-Foot, and Fine-Motor Coordination. At the beginning of each chapter a description of the main concepts and related subconcepts is included to help the reader understand the developmental levels and stages children pass through in their effort to establish these concepts as fundamental skills leading to more abstract generalizations and cognitive learning. Following this introduction, each chapter covers the acquisition and construction of the equipment and the objectives and suggested learning activities that will enhance perceptual-motor development.

Equipment listed within each chapter emphasizes a particular aspect of perceptual-motor development, although some individual pieces of equipment have several purposes, as can be observed in the list of objectives concerning their use.

The last four chapters are reference or resource sections. The first is a selected bibliography of books concerning learning disabilities and perceptual-motor learning. The reader may refer to these for an indepth knowledge of the theories, research, and approaches to learning disorders. The second of these chapters is an annotated film list concerning learning disabilities and perceptual-motor learning. A listing of diagnostic tests available for assessment purposes is the third chapter. Finally, selected commercial resource companies are listed to help the reader learn where equipment may be bought and at what cost in case that equipment cannot be adequately constructed through local resources.

2

Balance

Balance is the ability to maintain and/or regain one's body posture or position (8, 12, 27, 33). It is an essential component of all movement skills. Having good balance infers that one has internalized the relationship between the body's gravity center and it's support base. The body is balanced when it's gravity center is directly over the support base. Non-locomotor movements, such as body gestures with the arms and legs, occur in personal space when the center of gravity remains within the base of support. If the center of gravity falls outside the base of support, the body will move through general space. Efficient or poor movement is determined by the extent to which one can control and regain the loss of balance.

The body's ability to maintain and/or regain it's balance is monitored by certain sense organs (4, 36). The vestibular mechanism or semicircular canals and otoliths in the inner ear act as a center for balance. For this reason, tipping the body in various angles and directions and spinning techniques have been recommended for teaching children how to improve their stability. The muscles, tendons, and joints have somesthetic receptors, called proprioceptive end organs, that help a person gain the "feel" that he has about the position of his body in space. This is called the kinesthetic sense and can be taught by encouraging children to be more aware of their body and the relationships of its parts as move through space.

Visual perception also plays an important part in maintaining balance. The eyes help to determine the relative position of the body in space and without it, our awareness of our orientation in space is greatly altered. Children should be encouraged to balance their bodies in various situations, with their eyes open and then closed, to help orient their awareness of their body position in space with and without vision.

Balance has also been interpreted as a point of origin or "0" point for all other directions in space. It is from the center of gravity that the child orients himself to the directions of up, down, forward, backward, left, right, and the like (37). Thus the concept of balance is important as a reference point for spatial awareness.

Children should have experiences that require balancing the body in a static, stationary position and dynamically while involved in a variety of movement patterns. In general, they should learn to balance at a low level with several bases of support and move to more challenging experiences that involve fewer support bases at higher levels.

BALANCE PUZZLES

CONSTRUCTION:

The following balance puzzles can be made by drawing with a magic marker on 8½" x 11" sheets of paper or by making charts with poster board or cardboard. The sheets or charts may be laminated for a more permanent product. See figure 2.1.

OBJECTIVES:

1. To help children solve balance problems related to static balance.
2. To help children learn body awareness concepts.
3. To help children improve laterality and spatial orientation concepts.
4. To teach children better concepts of visual decoding, association, and memory.

ACTIVITIES:

1. Starting with the easier puzzles, hold up one chart at a time and ask the child to balance on the body parts, as shown in the puzzle. The child must balance with the correct parts (left or right) as well as in the same position, as shown in the chart, with no other body parts touching the floor.
2. Use the puzzles as part of an obstacle

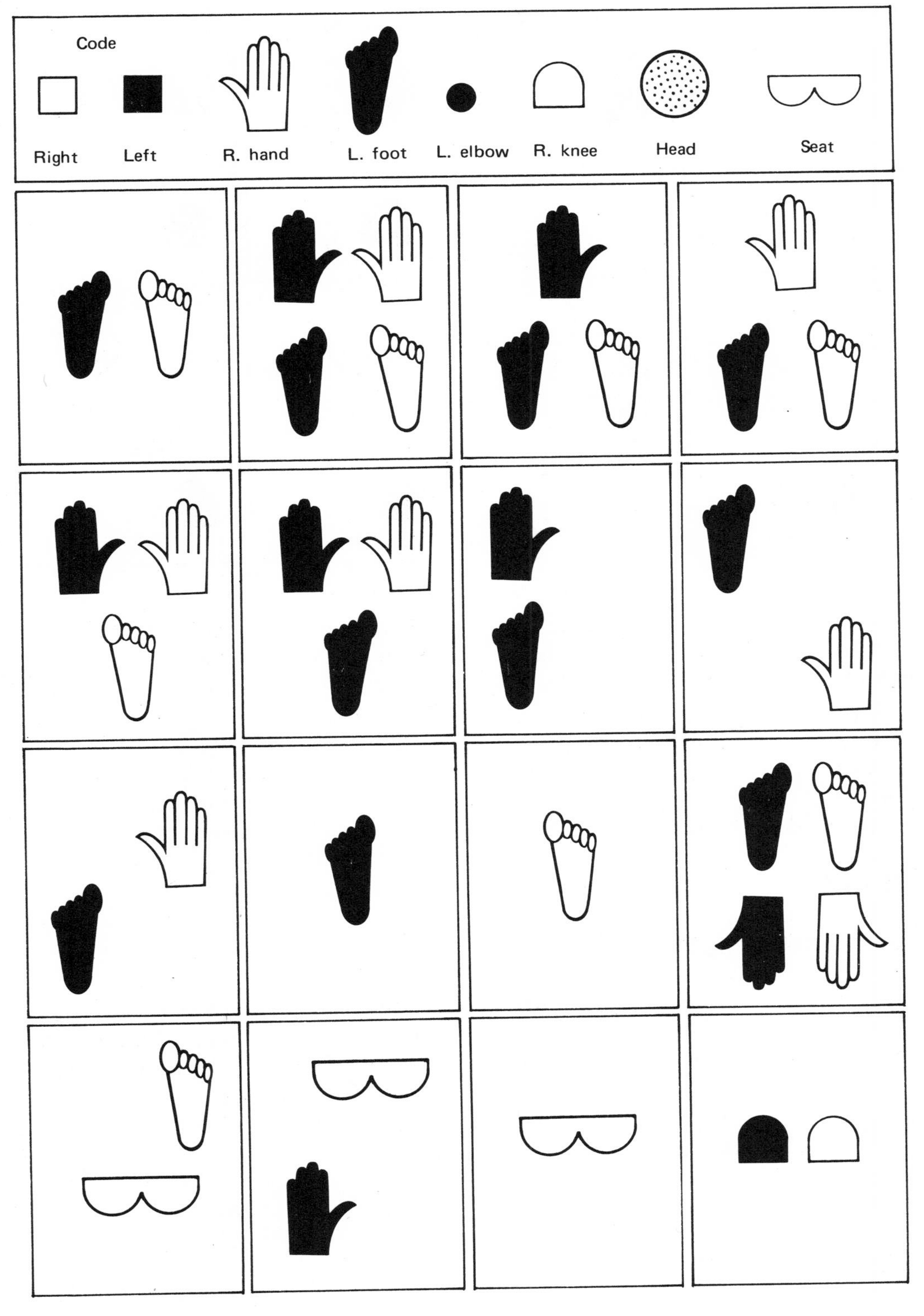

FIGURE 2.1. Balance puzzles.

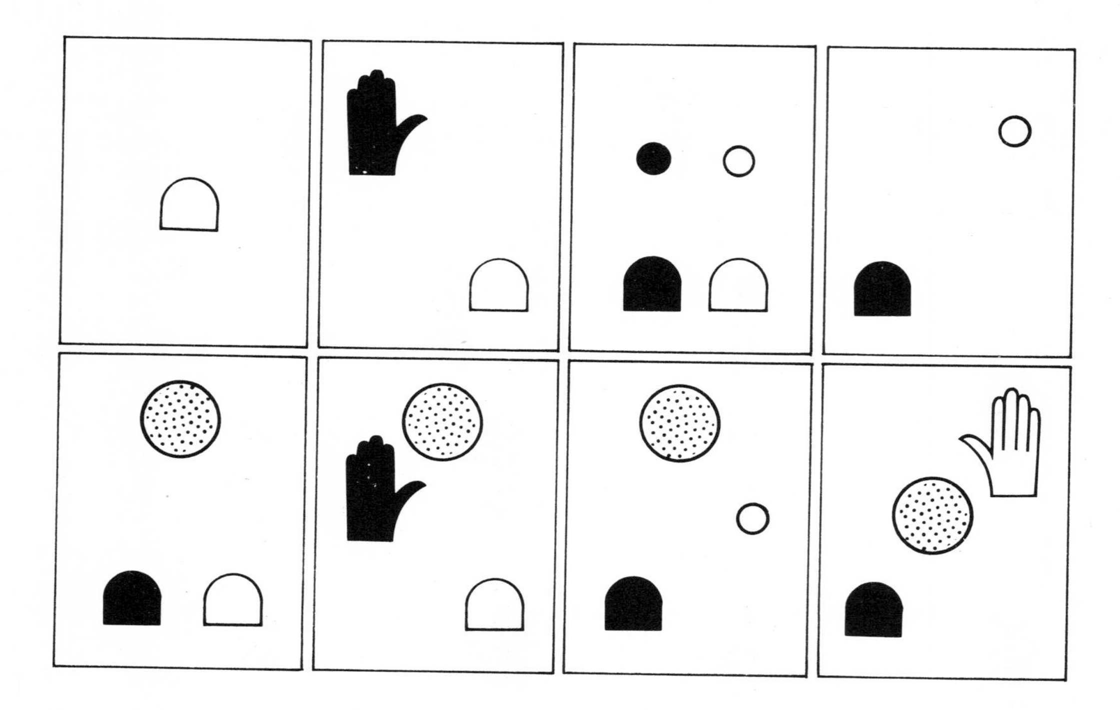

course. When the children arrive at one they must perform the designated balance task before moving on to the next obstacle.

3. Show the child a chart and then remove it from his vision. Ask the child to use his visual memory to assume the balance position.
4. Lay ten to twenty charts out on the floor and time the children to see how quickly they can assume each of the positions as they progress from one chart to another in sequence.
5. Show the child two, three, or four charts and then remove them from his vision. Ask the child to perform each of the balance positions in sequence from memory.

BALANCE BEAM

CONSTRUCTION:

The balance beam consists of a wooden plank, 8 to 12 feet long, and varnished to protect bare feet. Braces allow for additional height as the beam is inserted vertically for a 2 inch surface or horizontally for a 4 inch surface. Construct the brace by cutting a piece of 2 x 4 inches into three pieces 4 inches in length and attach them, as shown in the picture. Two 1 x 1 x 2 inch boards are attached to the inside of each brace. See Figure 2.2.

OBJECTIVES:

1. To develop better stability and balance patterns.
2. To increase the level of spatial awareness.

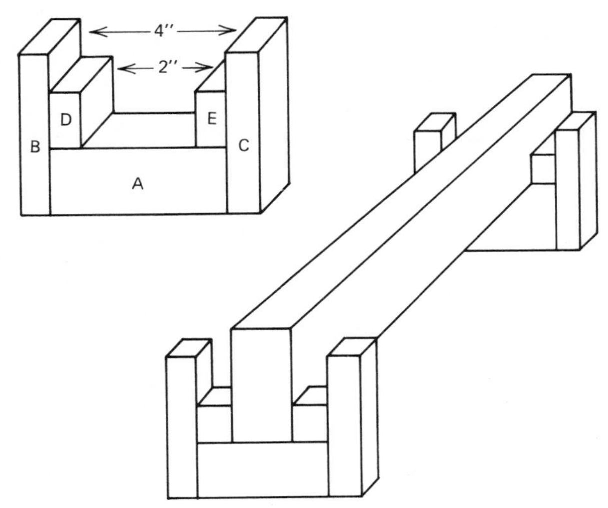

FIGURE 2.2. Balance beam.

3. To develop eye-foot and eye-hand coordination.
4. To increase tactile-kinesthetic awareness.

ACTIVITIES

1. Walk across the beam in forward, backward, or sideways directions.
2. Use other forms of locomotion to cross the beam. Examples may be sliding, skipping, galloping, or hopping.
3. Move across the beam with different bases of support including one, two, three, and four bases. Use different body parts to move across the beam.
4. Move across the beam while balancing an object, such as an eraser, on different body parts.
5. Turn around while walking across the beam.
6. Move across the beam and step over or under wands placed at various heights along the beam.
7. Stand on the floor and hop or jump from one side of the beam to the other. Hop and jump in forward, backward, and sideways directions.
8. Walk on the floor and place feet on alternating sides of the beam. Place the right foot on the left side of the beam and the left foot on the right side of the beam.
9. Roll a ball along the beam.
10. Bounce or dribble a ball as you walk across the beam.
11. Play catch with a partner as you walk across the beam.
12. Exchange places with a partner as you walk across the beam.

BALANCE BOARD

CONSTRUCTION:

The top of the balance board is made by cutting a piece of 3/4 inch plywood into a 16 inch square. The bottom of the base may be a 3, 4, or 5 inch square depending on the child's skill. Sink wood screws or nails to attach the top to the

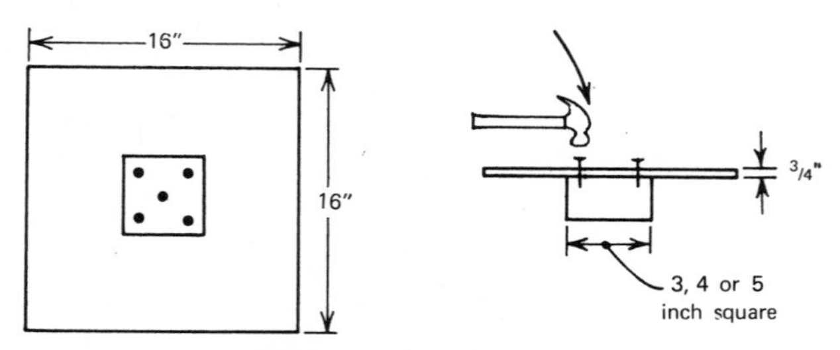

FIGURE 2.3. Balance board.

base. Rubber matting the size of the top, can be glued to the surface for added protection from slippage in case the child is off balance. See Figure 2.3.

OBJECTIVES:

1. To develop balance skills.
2. To increase the level of spatial awareness.
3. To increase the tactile-kinesthetic sense.
4. To increase eye-foot and eye-hand coordination.

ACTIVITIES:

1. Balance on the board and touch different body parts, including your knees, ankles, and toes.
2. Touch your various body parts with your eyes closed.
3. Try to stand on your tiptoes with your eyes open and closed.
4. Turn around while maintaining your balance on the board.
5. Force yourself to lose your balance and then regain it while moving in a forward, backward, or sideways direction. Place your feet in different stride positions as you do this. What is the best position for your feet as you tip in different directions?
6. Bounce a ball while you balance on the board.
7. Play catch with a partner as you balance on the board.
8. Jump rope while you balance on the board.
9. Assume different body positions with one, two, three, and four support bases as you balance on the board.

SPRINGBOARD

CONSTRUCTION:

Obtain a wooden board approximately 10' x 8" x 1" and two cement blocks. The board should be placed horizontally on the blocks with a slight overlap at the edges for added support.

FIGURE 2.4. **Springboard.**

OBJECTIVES:

1. To develop greater balancing skills.
2. To improve gross-motor coordination.
3. To increase spatial awareness.
4. To help overcome earthbound problems.

ACTIVITIES:

1. Have the child stand, with his feet together, in the center of the board. Have him jump several times in place, increasing his height as his skill increases.
2. Have the child step on and off the board several consecutive times. He can do this forward, backward, and to each side.
3. Tell the child to jump on the board a specific number of times and then to stop. Teach him to bend his knees, ankles, and hips to absorb the force and stop.
4. Use the board as a part of an obstacle course. For example, crawl under, jump over, walk across, and hop on the springboard.

* For additional ideas, refer to the balance beam and balance board activities in this chapter. (pp. 14-16)

ROCKER BOARDS

CONSTRUCTION:

A 1" x 6" x 2' board is needed for each rocker board. On the bottom of the board a 1 x 1" board can be placed as a track and boundary, as seen in Figure 2.5. A 2 to 5 inch cylinder should be used as the support or fulcrum for the rocker board.

FIGURE 2.5. **Children can balance on the rocker boards.**

OBJECTIVES:

1. To help children learn static-balance concepts.
2. To help children develop improved concepts of laterality, directionality, and spatial awareness.
3. To help children cross the midline of their bodies.

ACTIVITIES:

1. Have the child roll from side to side and touch the right and left ends to the floor alternately while maintaining control of his body. Have the child use his arms for more control.
2. Have the child try to maintain balance in the middle without either end touching the floor for as long as possible.
3. Have the child cross his legs so that the right leg will be forced to balance on the left side and vice versa.
4. Have the child bounce a ball or perform various visual pursuit tasks while balancing on the rocker board.

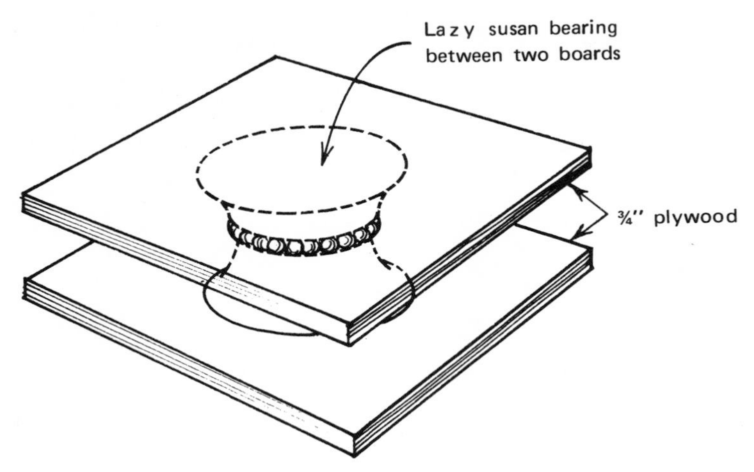

FIGURE 2.6. **Twister board.**

TWISTER BOARD

CONSTRUCTION:

A twister board can be made by using two pieces of 3/4" x 10" x 10" plywood separated by a lazy susan bearing. Attach the two pieces of plywood to the bearing with wood screws. The bearing allows the bottom piece of plywood to remain stationary while the top piece of plywood moves around. See Figure 2.6.

OBJECTIVES:

1. To improve balancing skills.
2. To develop gross-motor coordination.
3. To increase spatial awareness.
4. To assist the child in learning how to cross the midline of his body.

ACTIVITIES

1. The child should place his feet shoulder-width apart on the board and bend his knees slightly. Twist on the board by moving both of his arms to one side or the other.
2. Have the child attempt quarter, half, and full turns by moving only one arm. Start the arm by reaching across the body's midline. Then push the arm quickly out to the side while the trunk of the body moves in the opposite direction.
3. The child can sit or kneel on the board and twist.

4. Begin standing, twist to a squat position, and return to an upright position.
5. Have the child develop a pattern or sequence of different twists in different body positions. Allow him to perform the pattern to a rhythmical beat.

TIN CAN STILTS

CONSTRUCTION

When preparing cans for walking and balancing activities, the number 7 size industrial food cans or large coffee cans work best. Punch a hole on each side of the can on the end where the lid is still intact. Insert one end of a jump rope in each hole and tie a knot in each end so that the rope will not pull back through the holes. The loop of the rope should be about hip-high. To prevent wear on the rope, wrap tape around the rope where it passes through the holes. Place the plastic lid from the coffee can on the open end of the can or tape the edge so that the floor will not be scratched. Provide two cans for each child. See Figure 2.7.

OBJECTIVES:

1. To increase a child's balancing skills.
2. To improve eye-hand and eye-foot coordination.

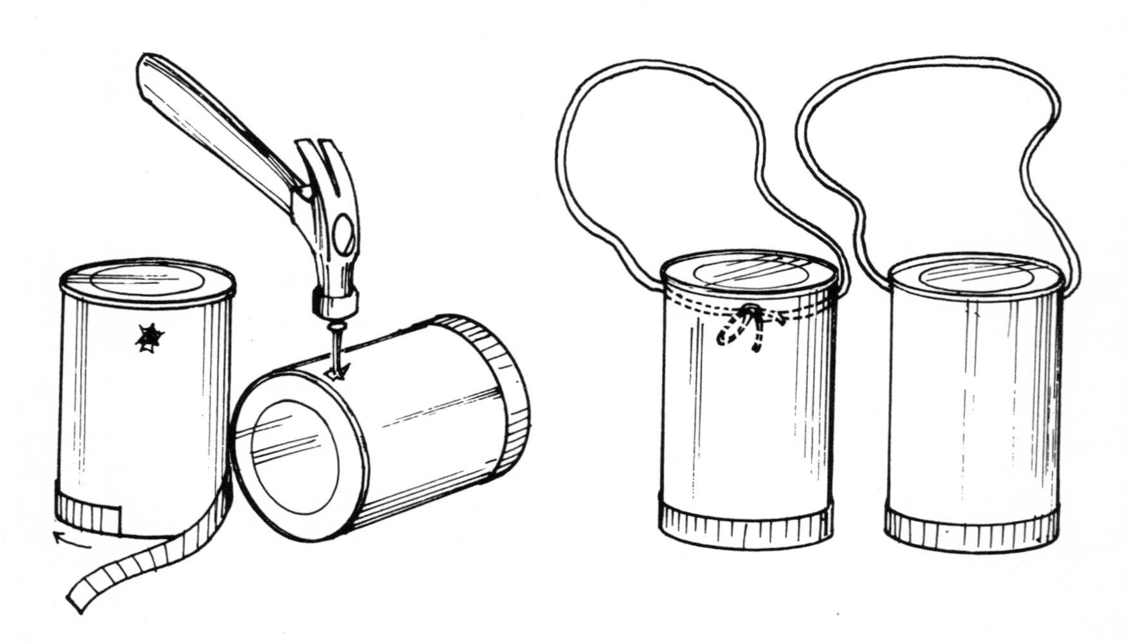

FIGURE 2.7. Tin can stilts.

3. To develop gross-motor coordination and agility.
4. To increase a child's awareness of laterality and directionality.

ACTIVITIES

1. With one can under each foot, while grasping the cords in each hand and holding the cans tightly to the feet, the child should try to walk forward, backward, and sideways using the cans as stilts.
2. Try to perform various hopping or jumping activities while using the cans as stilts.
3. Have the children move through an obstacle course or perform various relay activities while using the tin can stilts.
    a. Walk forward on the stilts.
    b. Walk backward on the stilts.
    c. Walk sideways on the stilts.

STILTS

CONSTRUCTION:

Purchase a 8' x 4" x 2" pine or yellow poplar board from a local lumber company. Cut 1 foot off the end of the 8-foot board. Take the 1 foot board and cut it in half. Take one half and cut it diagonally to make the steps. Cut the 7' x 4" x 2" board vertically to make two poles. Attach the steps with nails to the side of the poles at the desired height. Tape the bottom of the stilts to prevent slippage and the center of the stilts to prevent splinters. See Figure 2.8.

OBJECTIVES

1. To increase the level of balancing skills.
2. To improve the eye-foot and eye-hand coordination.
3. To develop spatial awareness.

ACTIVITIES:

1. Methods of getting on the stilts.
    a. Lay the stilts on the floor. Stand at the top of the stilts. Pick up the stilts and place them under the armpits with the arms around the front of the poles. Place one foot on one of the steps, push up with the free leg, and place the free leg on the step of the second stilt.
    b. Hold onto the stilts, as in the previous procedures. Mount the stilts from an elevated position, such as a chair.
2. Walk forward, backwards, and sideways with the stilts.
3. Create an obstacle course. Give students an opportunity to step over, under, around, and through objects.
4. Place hands on the top of the stilts and walk.

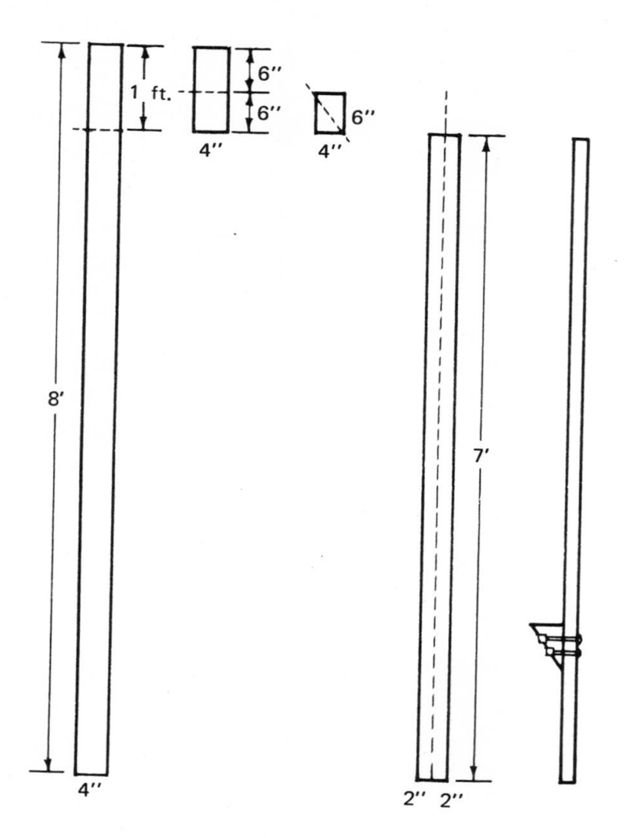

FIGURE 2.8. Stilts.

5. Try to jump or hop while on the stilts.
6. Try to walk up steps while on the stilts.
7. Stand on one stilt, lift the other leg up, and swing it around.

BARRELS

CONSTRUCTION:

The barrels are the center cores of large rolls of papers and can be obtained through a paper mill or cardboard company. The barrels need no further construction unless it is desired to paint them with letters or geometric forms. See Figure 2.9.

OBJECTIVES:

1. To help children develop balance.
2. To develop gross-motor coordination and agility.
3. To develop spatial awareness and directionality.

FIGURE 2.9. Balance on the barrel.

ACTIVITIES:

1. Different ways to push the barrel.
    a. Push the barrel with just the right hand or just the left hand.
    b. Push the barrel with both hands.
    c. Push the barrel with both feet.
    d. Push the barrel with one foot.
    e. Push the barrel with the other foot.
2. Different ways to balance on the barrels.
    a. Balance on the barrel with both feet.
    b. Walk on the barrel and move forward.
    c. Walk on the barrel and move backward.
    d. Jump on and off the barrel while it is stationary.
    e. While the barrel is moving forward, bring it to a stop, jump and turn around.
    f. Jump rope while on the barrel.
    g. Bounce a ball while on the barrel.
    h. Work out partner routines on the barrel.
    i. Lie on stomach or back and balance on top of the barrel.
3. Ways to use the barrels as they stand on end.
    a. Use the barrels for target toss with balls, beanbags, and other objects.
    b. Line the barrels in a row; use as barriers for zigzag running, skipping, hopping, dribbling a ball, and other challenges.
    c. Use the barrels for obstacle courses.
    d. With ends open, the children can crawl through the barrels.

TRAMPOLINE

CONSTRUCTION:

The trampoline can be purchased commercially. If funds are low, an old mattress or box springs serve the same purpose and may be preferred if the height of the trampoline is a threat to the child. See Figure 2.10.

FIGURE 2.10. Trampoline.

OBJECTIVES:

1. To develop better balancing skills.
2. To increase the level of gross-motor coordination.
3. To develop spatial awareness, laterality and directionality.
4. To increase body awareness.

ACTIVITIES:

1. *Introduction to trampoline.* Allow the child to become familiar with the trampoline. While standing on the floor, have him:
   a. crawl under the trampoline.
   b. walk around the trampoline.

c. bounce the bed of the trampoline with his hands.

On the bed of the trampoline, have the child:

a. crawl around the edges and then into the center of the trampoline. This helps him understand that the bed is softest in the middle.
b. roll from side to side. This helps the child discover the size of the trampoline bed.
c. walk around the edges and across the middle. In this activity, the child can feel the "give" in the bed.

2. *Bouncing.* When training a child on the trampoline, bouncing activities should be considered the basic or primary skill.
   a. Bounce the child while he is lying in a prone and supine position.
   b. Have the child initiate his own bounce while in the prone and supine position.
   c. Repeat steps *a* and *b* when the child is in a hands and knees position.
   d. Have the child stand in the middle of the bed and learn to bounce by bending his knees and thrusting against the bed with his feet. If he has difficulty, support him through the activity by holding his hands or by letting him use your arm for support.
   e. Instruct the child that he can stop bouncing by quickly bending his knees. This halts the motion of the trampoline bed. Have the child practice "1-2-3, Stop bouncing!" several times. The child should achieve this skill before being allowed to perform without a rope or safety belt. (Romper Trampoline.)
   f. Instruct the child to bounce on one foot. Have him repeat this activity on the opposite foot.
   g. Instruct the child to bounce to a 1-1 rhythm (bounce on the right foot one time, then left foot one time). Repeat ten times.
      (1) 2-2 rhythm.
      (2) 3-3 rhythm.
   h. Use combination of rhythms, such as 2-3, 4-1, 3-2, 1-2 with the teacher holding up his fingers to indicate the number of bounces on each foot. The number of fingers should be altered while the child is bouncing on the opposite foot.
   i. Repeat moving from side to side.
   j. Place numbers or various symbols at given points on the trampoline bed. When a number or symbol is called, the child must jump to it.
   k. Have the child do "jumping jacks" on the trampoline. If he has difficulty, tell him to use only his legs. If he still has difficulty, move the child to the edge of the trampoline. Have him lean over you and place his hand on your back. Start him bouncing by holding his feet and moving them apart and together.
   l. The student assumes a forward stride position. He alternates the position of his feet with each bounce.
   m. Have the child jump forward and backward. Tell him to say the direction he jumps.
3. As the child's skill level increases, he can attempt stunts. Various stunts include the knee drop, seat drop, hands and knees drop, front drop, back drop, and combinations of these. Full, half, and quarter turns may also be included.*
4. After the child has developed the basic skills on the trampoline, have him jump rope doing skills previously listed.

* For a more detailed account of trampoline skills, consult the Nissen Gymnastics Company, Cedar Rapids, Iowa.

FIGURE 2.11. Crawl through the ladder.

LADDER

CONSTRUCTION:

There is no specific construction for the ladder; one could be made easily. To construct one, purchase two 12' x 2" x 2" poles and the desired number of rungs, 2' x 2" x 2". Nail the rungs to the poles at the desired spacing. Whether the ladder is constructed or ready-made, sand it down and varnish it to protect the children from splinters. See Figure 2.11.

OBJECTIVES:

1. To develop balance skills.
2. To increase eye-foot and eye-hand coordination.
3. To increase gross-motor coordination and agility.
4. To increase spatial awareness, laterality, and directionality.

ACTIVITIES:

1. *Walking.* Looking straight ahead, the child walks the length of the ladder:
   a. forward in the spaces.
   b. backward in the spaces.
   c. sideways in the spaces. Walking sideways may be done by:
      (1) leading with the left foot.
      (2) leading with the right foot.
      (3) continually crossing the lead foot in front.

(4) continually crossing the lead foot in back.

d. on the rungs (forward, backward, sideward).

e. on all fours, first walking on the rungs, then on the rails (the sides of the ladder).

f. on the rails.

g. as in step *e*, but the front of the body faces upward (crab walk position).

2. *Jumping.* The child moves the length of the ladder, jumping (feet together):
   a. forward in the spaces.
   b. forward in every other space.
   c. backward in the spaces.
   d. sideward in the spaces.
3. *Hopping.* The child moves the length of the ladder (on one foot):
   a. forward in the spaces.
   b. sideward in the spaces.
   c. forward for three spaces on one foot, and then changing and hopping on the other foot for the next three spaces, and so on.
4. The child stoops down at the end of the ladder, places both hands on the second rung from the end, and jumps into the first space. He moves his hands to the next rung, and jumps into the second space. He continues this pattern, moving to the other end of the ladder.
5. The ladder is turned on its side, and to secure it, the teacher sits on the top side. Then the child crawls in and out the spaces.
6. Use the ladder as part of an obstacle course for the children to crawl under, over, and so on.

TIRES

CONSTRUCTION:

There is no construction necessary. Old automobile tires can be obtained at any service station or junkyard. See Figure 2.12.

OBJECTIVES

1. To help develop better balance.
2. To increase spatial awareness.
3. To develop eye-hand and eye-foot coordination.
4. To help children develop manipulation skills.

ACTIVITIES:

1. Crawl through one tire.
2. Crawl through two or more tires while they are held upright.
3. Roll the tire.
4. Roll the tire; run and jump over the rolling tire; turn and stop the rolling tire.
5. Roll the tire; run in front of the tire and stop it.
6. Roll and catch the tire with a partner.
7. Spin the tire like a coin: as it settles to

FIGURE 2.12. Tire activity.

the floor, run around it as many times as possible before it stops.

8. Bounce on the tire; use it as a mini-trampoline.
9. Bounce from tire to tire in a pattern.
10. Hop in and out of individual tires; set the tires in a pattern and hop in and out of tires. (Line, circle, zigzag, etc.)
11. Extend the body *through* the tire and roll, over and over.
12. Sit on the tire (upright) and balance.
13. Use the tire as a weight; pick it up and raise as high as possible.
14. Use the tire as a target for beanbag toss, yarn-ball toss or playground-ball toss (stationary).
15. Use the tire as a target (moving) by hanging it from a basket support with a rope; as the tire swings, throw beanbags or balls through the tire. Make up a game.
16. Many of the listed activities can be used in *obstacle courses*.
17. Various playground activities can be developed.

LARGE INNER TUBES (CAR, TRACTOR, OR TRUCK)

CONSTRUCTION:

Large inner tubes can be obtained from junkyards, tire dealers, farm-implement stores, or farms. Any large size tube could be used if it holds air. Patches might be necessary. Caution should be taken by covering the valve stem so that it does not scratch or puncture a child. See Figure 2.13.

FIGURE 2.13. Large inner tubes.

OBJECTIVES:

1. To help children develop balancing skills.
2. To help children develop agility and coordination.
3. To help children develop spatial awareness.

ACTIVITIES:

1. Walk on the tube while it is lying on the floor.
2. Jump onto the inner tube and rebound off. Make ¼, ½, and full turns as you rebound off the tube.
3. Roll the inner tube across the floor and try to run or dive through it as it is moving.
4. Tie a rope to the tube and pull a partner as he sits on the tube.
5. Use the tube for a 2-, 4-, or 6-person tug-of-war.

* For additional ideas, refer to the tire activities in this chapter (pp. 27-28).

CARPET SAMPLES

CONSTRUCTION:

Carpet samples can be obtained from carpet stores. The carpet should be cut in 12 or 16 inch squares. If remnants are the only samples available, simply cut them to size and spray the edges with hair spray to prevent fraying. For best results place the carpet side nearest the floor. See Figure 2.14.

FIGURE 2.14. Carpet activities.

OBJECTIVES:

1. To further develop balancing skills.
2. To increase gross-motor coordination.
3. To help children develop a sense of rhythm.

ACTIVITIES:

1. Individual.
   a. Skate or ski by putting one rug under each foot. Keep feet well apart for balance. Use arms for a skiing or skating motion.
   b. Place knees in the middle of the carpet and pull forward with hands and arms.
   c. Sit in the middle of the carpet and push with the feet and legs. Hands may also be used.
   d. Sit in the middle of the carpet and pull with the feet and legs.
   e. Use a body twist to move the carpet. By keeping the feet in the middle of the carpet, the student can twist in many different ways. Using music makes this activity even more enjoyable.
   f. Assume a push-up position, with hands on the floor and feet on the carpet. Move forward by "walking" the hands as the feet stay on the carpet. This can be called the seal crawl.
   g. Use two carpets and move by placing the hands on one carpet and knees on the other.
   h. Lie down with back flat on the carpet. Use feet and legs to push the carpet.
   i. With both feet on the carpet, use a jumping motion to move forward or backward. Grip the carpet with the feet.
   j. Allow the children to have a "creative" time where they can explore many additional movements.
2. Partners
   a. With one student seated (crossed legs), have the partner carefully pull his mate by his hands. Insist that the students hold both hands and stay on the carpet.
   b. Tag and games of chase may be used with carpets.
   c. One child sits with his feet on one carpet and his seat on the other. His partner uses a jump rope (or other rope) to pull his mate. Best leverage is acquired with the rope around the waist of the person pulling.
   d. Try all types of races and relays with the carpets.
   e. Small groups of children may work together in forming patterns, shapes, designs, and movements with the carpets.
3. Other Uses or Suggestions:
   a. Use the carpets as targets for beanbag toss.
   b. Cut the carpets into geometric shapes and use with correlation activities.
   c. Use rubber-backed carpet for goals, bases, or markers.
   d. Stress safety at all times.
   e. Best movement occurs with the carpet-side *down*.
   f. Try to keep the children on the carpet at all times.

SCOOTER BOARDS

CONSTRUCTION:

One 12" x 12" x 3/4" piece of plywood with two good sides and four good quality casters with rubber wheels are needed for each scooter board. Round off the corners of the 12" by 12" x 1" plywood so that the edges become smooth. Mount one caster on each of the four corners with wood screws. If more than one

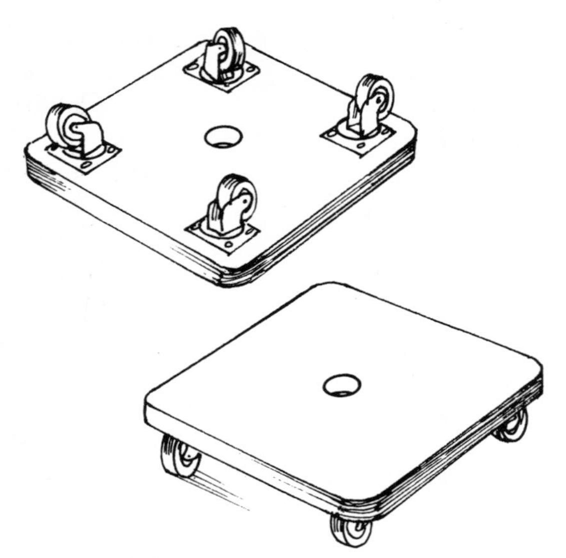

FIGURE 2.15. Scooter boards.

scooter board is constructed, storage is simple if a 1 inch hole is drilled in the center of the plywood. Then they can be stacked vertically by placing them on a dowel. See Figure 2.15.

OBJECTIVES:

1. To increase the level of balancing skill.
2. To increase the gross-motor development of the upper arm and trunk strength in children.
3. To increase spatial awareness.

ACTIVITIES:

1. Individual activities.
   a. Kneel on the scooter board and push yourself around the gymnasium or obstacle course with your arms.
   b. Lie down on your stomach on the scooter board and push yourself around the room with your arms.
   c. Sit down on the scooter board and push yourself around the room with your legs.
   d. Utilize the above activities in relays.
   e. Modify the games of basketball, soccer, or other games, by using the scooters as the source of locomotion.
2. Partner activities.
   a. While one child sits or kneels on the scooter, have a partner push him about the room or through an obstacle course by placing his hands on the shoulders of the child sitting down.
   b. Allow one child to pull another about the room by having one child hold tightly to the ends of a jump

rope, as he sits on the scooter and by having a partner put the middle of the rope around his waist, while he runs around the room.

c. Perform the above activities in relays.

PLUNGER

CONSTRUCTION:

A household plunger, thumbtacks, small plastic toys, and small magnets are needed to construct this piece of equipment. Stick three or four thumbtacks into the handle of the plunger. Space them apart evenly. Glue a magnet to each of the small plastic toys. See Figure 2.16.

FIGURE 2.16. Jump and reach for the toy on the plunger.

OBJECTIVES:

1. To help earthbound children learn to leave the ground.
2. To help children develop gross-motor coordination.
3. To increase a child's eye-hand coordination.

ACTIVITIES:

1. Attach the plunger to a wall, just above the child's reach. Place a plastic toy with it's magnet at each of the thumbacks. Tell the child to jump up and grab each of the toys off of the handle of the plunger.

* Variations of this activity are to use clothes lines and pins or balloons. String a clothes line over the child's head so that it is just beyond his reach. Attach several clothes pins. Have the child jump up and try to remove each of the clothes pins. When using the balloon, suspend it over the child's head, just beyond his reach. Have him jump up to hit it or try to pop it with a pin.

3

Body Image

Body image is a complex factor that initially involves a person's ability to identify his body parts and know how each can be used in interacting with the environment (5, 7, 11, 12, 21, 27, 28, 33, 35). Gradually, these abilities enable the child to identify a personal self-image or self-concept relating to his interpretation of self-worth (success-failure ratio) and how the individual thinks others view him. It is estimated that, even while in the womb, a fetus's movements help to form the first perceptions of the body and it's capacities for motion (12). As a small infant and during early childhood, a person begins to identify body parts and learns how each can be used to interact with the environment. After the child learns the gross body parts, the more abstract parts are gradually accommodated. The small child also learns that the body has two sides, though it may not be until the ages of six or seven before a child can accurately identify right and left.

Learning one's body parts and knowing how to function efficiently in the environment is extremely important in the development of the self-concept. In early childhood, peer relationships are often affected by how well one can perform playskills. Most young children find it important that they know the degree to which they can perform a task. The feeling they have toward their performance abilities and how others accept this activity exerts a great deal of positive or negative influence upon their total self-concept. The child with a low concept of his performance must be placed in a situation where success can be achieved. As the child's self-concept is improved, more challenging situations may be introduced. When the success ratio is greater than the failure ratio, the child will probably adopt the attitude of "I'll try" or "I can." For this reason, teachers and parents must insure that all children receive a variety of movement-pattern experiences aimed at each child's ability level. The successful child who identifies his body parts and knows how to move efficiently will have the foundation needed for a positive self-concept.

PIPE CLEANER FIGURES

CONSTRUCTION:

Mold pipe cleaners into the shape of a human form, as shown in Figure 3.1. Twist several pipe cleaners together to achieve more stability and greater dimension in the trunk and limb areas. Different colors can be used to identify body parts or aspects of laterality (right side is made of red pipe cleaners; left side is made of blue pipe cleaners).

OBJECTIVES:

1. To enhance concepts of body image and body awareness in children.

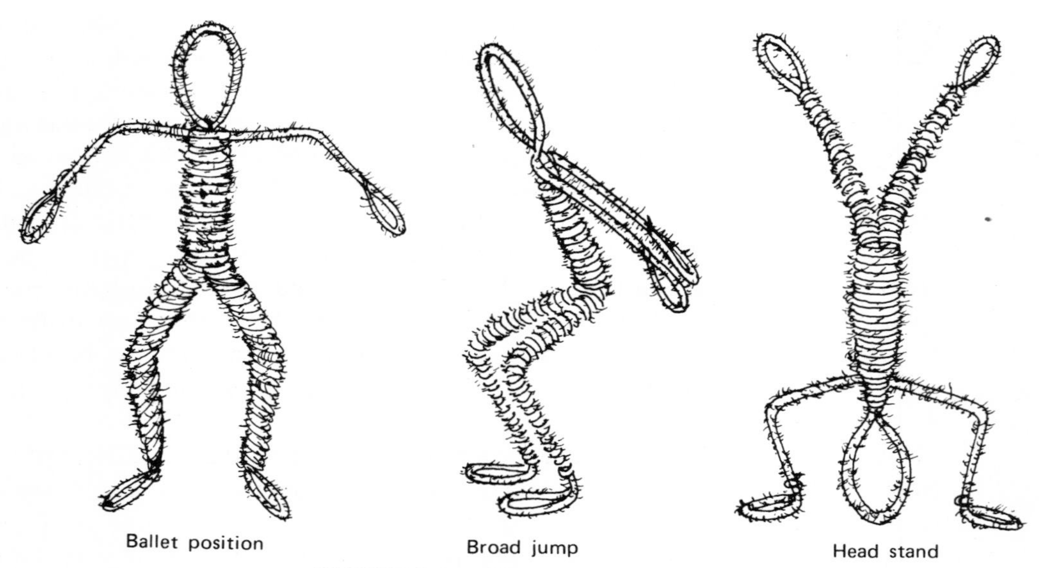

FIGURE 3.1. Pipe cleaner figures.

2. To develop concepts of laterality and directionality.
3. To provide children with opportunities to combine fine-motor and gross-motor coordination.

ACTIVITIES:

1. Develop a series of pictures with human forms in various poses. Have the child bend the pipe cleaner figure into these different poses.
2. Have the child assume the same pose as the picture and pipe cleaner figure.
3. Have one child assume a body pose and have another child manipulate the pipe cleaner figure into the same pose.
4. Have the child assume the exact opposite position of the pipe cleaner figure.

PUZZLES OR MANIKINS

CONSTRUCTION:

Human body puzzles can be made in several ways. For example, cut out the shape of a human form from a piece of press board. Paint each of the body parts on the figure. With a jig saw, cut the puzzle into pieces—head, arms, hands, trunk, legs, and feet. A second type of puzzle can be made by using pieces of cloth and sandpaper. Cut different textures of cloth into several body parts. Paste these parts onto cardboard for backing. Use the sandpaper to identify some of the parts. A third puzzle can be made by cutting the shape of a body out of a piece of cardboard. On the cardboard, draw in the details of the head, feet, and hands with chalk or crayon. Then, cut out clothes, shoes, and other garments from magazine advertisements for the model to wear. The children can also make designs of their own clothes they wish the models to wear. Fabricate a story in which the models role play. Have the children manipulate the models and perform any necessary clothes changes. Have the children make up their own stories. A wooden maniken also allows the children to perform role-playing experiences. The authors have discovered that it is best to purchase one of these from an art supply store. Homemade ones are not durable and do not perform the movements of the limbs correctly. See Figures 3.2 and 3.3.

OBJECTIVES

1. To develop body awareness and body image.
2. To develop an awareness of space.
3. To develop eye-hand coordination.

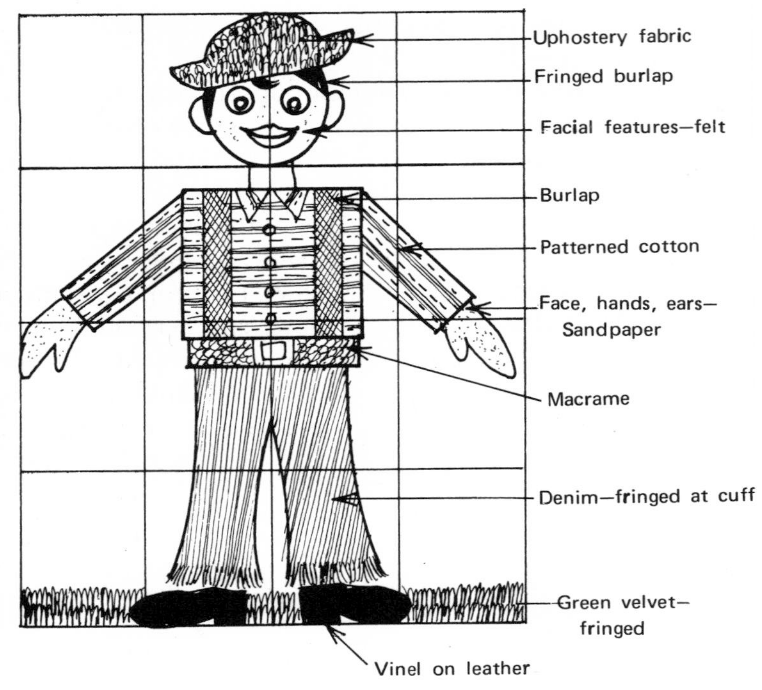

FIGURE 3.2. Puzzle.

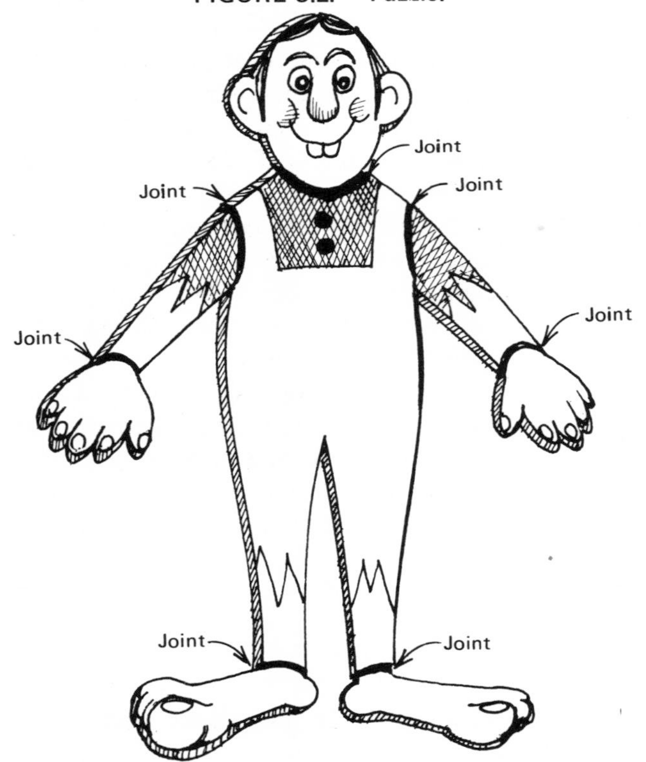

FIGURE 3.3. Manikin.

ACTIVITIES:

1. Put the human figure puzzle together. Identify each of the body parts.
2. Make your own clothes or cut pictures of clothes from magazines and place them on your models.
3. Manipulate the parts of the puzzle to create a body pose. Can you make your body assume the same pose?

CHALKBOARD AND MIRROR

CONSTRUCTION:

Most schools have adequate chalkboard, but, if necessary, an 8 x 4 ' sheet of masonite painted with chalkboard paint makes a good substitute. Mirrors should be full-length. Sometimes they can be acquired when parents purchase a new bedroom suite and are willing to donate the large mirror from their old one. See Figure 3.4.

OBJECTIVES:

1. To develop body image.
2. To develop tactile and kinesthetic awareness.

FIGURE 3.4. **Chalkboard or mirror activities.**

3. To develop an awareness of laterality and directionality.
4. To help the child learn the concepts of auditory and tactile decoding, sequencing, and memory.

ACTIVITIES:

1. Use chalk or shoe polish and have one child draw the outline of another on the chalkboard or mirror. Have the children draw and complete the picture by filling in all of the human parts (eyes, ears, fingers, mouth, etc.). Have them name each of these parts.
2. Do the above activity with the children assuming different poses. Have the children point to the left arm, right ear, left leg, and so on.
3. After outlining the child's body on the board or mirror, place nine *X*'s, three in each of three rows, as shown in the picture. Identify where each of the *X*'s is located on the back of the child. Then strategically touch the child's back at the location of one of the *X*'s. The child's task is to identify the *X* touched on his back by locating the same *X* on the board or mirror. (See Figure 3.4.)
4. Tap out a pattern from one spot to another, starting with two locations and building to three or more. Repeat in irregular rhythm. Have the child identify the pattern by sequentially and rhythmically touching the *X*'s on the board.
5. While still using the above *X*'s, trace a line from one dot to the next on a child's back. Have him duplicate on the chalkboard, the line he felt on his back. Start with horizontal lines first, then vertical, and oblique. Make sure the child duplicates all of the aspects of the line he felt—the direction, speed, pressure, and position.
6. Trace various geometric shapes on the child's back. Help him to recognize differences in positions on his back, size, pressure, speed, and direction by seeing that he traces appropriately these aspects of the shapes onto the chalkboard.
7. Using a buzzer behind the child's back, make intermittent short and long sounds. Have him draw short and long lines on the chalkboard. Get him to be aware of the length of time of the sound and relate it to the length of the line he draws. Start with single sounds of different lengths aand move to groupings of sounds for him to replicate on the chalkboard.
8. Have the child decode from the chalkboard by using a buzzer to sound out the series of dots and dashes from step 7.

NEWSPAPER OR LARGE PIECES OF PAPER

CONSTRUCTION:

There is no construction involved in this experience on the part of the teacher. All the teacher needs to do is provide the raw materials—newspaper, large rows of paper, crayons, finger paints, chalk, and/or texture materials.

OBJECTIVES

1. To develop awareness of one's body parts.
2. To develop flowing movements and freedom in the child's drawing skills.
3. To develop eye-hand coordination.
4. To develop an awareness of the environment through haptic experiences involving, color, shape, texture, smell, and the like.

ACTIVITIES

1. Lie down on the floor on a sheet of paper and have someone draw the outline of your body with one of the art materials.
2. Fill in and identify the various parts of your body.
3. Color in the various parts of your body according to the color of your hair, shirt, shoes, and so on.
4. By using a projector and shining the light on a wall, assume your favorite sports position (batting a ball, jumping rope, etc.) so that your shadow can be drawn on a sheet of paper. Fill in your body parts, as you did in the latter activities. See Figure 3.5, p. 42
5. Perform the above experiences with finger paints by having the children draw pictures or simple designs. Use basic finger paints or add variability to paints, such as spice smells, tint bases to change the colors, powder or oil to make the paint slippery or greasy, and textures, such as raisins or coconut for a lumpy texture. The variability introduced to the paints will make the child more discriminatory in terms of color, smell, and texture. This, in turn, will help the child become more aware of his body and senses.
6. Have the child draw simple designs on a large sheet of paper with the emphasis on smooth and free-flowing lines. On a second sheet of paper, have the child draw angular lines with abrupt changes of direction. Play smooth and free-flowing string or symphony music in contrast to percussive music for increased awareness. Have the children create a body shape or design similar to the design they have drawn. Have the children reproduce their design by walking the pattern out on the floor.

FELT BOARD

CONSTRUCTION:

Glue or staple a large piece of felt material to a piece of cardboard, masonite, plywood, or the like. The board should be at least 2' square or more in dimension. The board may have a plain background or consist of different colors as in an environmental felt board seen in the picture. Cut various pieces of felt into different shapes, forms, letters, numbers, and the like to be placed on the board. See Figures 3.6, 3.7, and 3.8.

FIGURE 3.5. **Draw a picture of a sports pose.**

OBJECTIVES:

1. To develop a better concept of body image and self awareness.
2. To help children learn the concepts of laterality and directionality.
3. To develop better form perception.
4. To develop fine-motor coordination.

ACTIVITIES:

1. Cut pieces of felt into images representing the right and left hands. Place them in different positions on the felt board. Have the child identify which hands are right and/or left.
2. Have the child put the pieces of felt together in the shape of a person.
3. Perform various experiences with letters, numbers, and geometric shapes. Make words from letters, perform math problems, and group all geometric shapes of the same kind.
4. Place all of the animals in their proper domain on the environmental felt board (for example, fish live in water, turtles live in water and on land, and birds live in the air and on land.)

FIGURE 3.6. Geometric person.

FIGURE 3.7. Right-left hand?

Sample of pieces to apply:

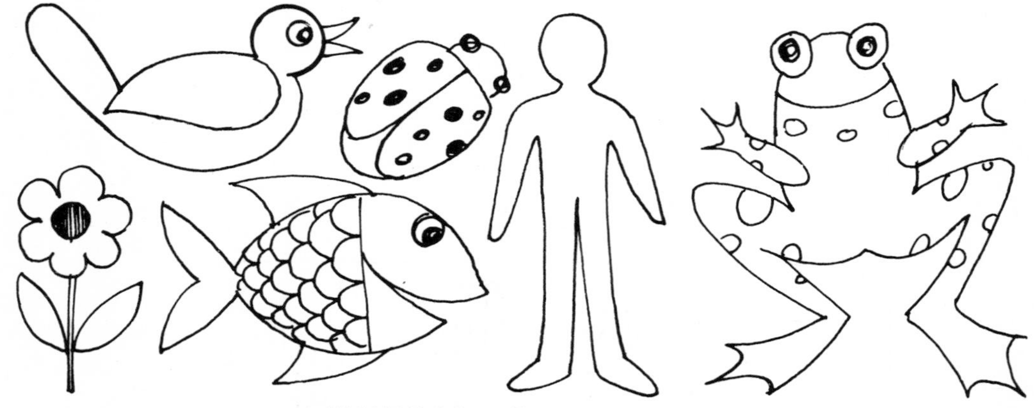

FIGURE 3.8. Environmental felt board.

4

Spatial Awareness

Spatial awareness is the collective term referring to the child's ability to learn the concepts of laterality, directionality, position in space, and spatial relationships (19, 27, 35). During early childhood, a person becomes internally aware of the body's two sides. This internal awareness or laterality extends to external space, as the child learns the directions of right, left, forward, backward, up, and down. Soon the child learns to use these directionality spatial concepts to locate objects in the environment with reference to his position. This concept has been termed egocentric localization or position in space. As the child learns to grasp more abstract concepts, two or more objects can be located with reference to each other. This is termed subjective localization or spatial relationships and is the most complex level of a person's ability to locate objects in the environment.

These concepts have been grouped together because they serve as building blocks; one concept builds upon another until the highest level is achieved (15, 25, 27, 41). It is important for children to learn these successive concepts through concrete practical experiences that will enable them to more easily grasp the abstract spatial terms. It is the writers' contention that the child who has experienced movement to the right, left, up, down, forward, backward, over, under, around, through, and other spatial terms will more readily understand the concepts than will children who have not had these learning experiences. It is also hypothesized that gross-motor experiences will have transfer into readiness skills for letter recognition, as the only difference between the letters of "b" and "d" is the spatial dimension of right and left. The authors are aware, however, that other factors such as midline, reversals, and visual perception may affect a child's performance in reading readiness.

CHARTS AND TRANSPARENCIES

CONSTRUCTION:

Make charts or transparancies with designs of arrows, triangles, and/or fingers drawn on them, as seen in Figure 4.1. The chart should be made of cardboard 2'x 3' in dimension. Use a felt tip marker to draw the designs on the chart in a left to right and top to bottom orientation. Use a grease pencil to make the same designs on the transparencies. See Figure 4.1.

OBJECTIVES:

1. To help children develop laterality and directionality concepts.
2. To develop a system of organization that proceeds from left to right and from top to bottom.
3. To increase a child's awareness of rhythm.

ACTIVITIES:

1. Place the chart in front of the child or show the transparency on a screen in front of the child. Choose the arrow design first. If the child doesn't know left and right, have him give the directions of a chosen row by saying, "Up, down, and side." After doing a single row, progress through the whole chart. Then use the triangle chart where the apexes point in the four directions.
2. Once the child is ready to learn left and right, have him perform the same previous activities by calling out, "Right, left, up, or down," as he progresses through the charts.
3. Perform these same activities to the rhythm of a metronome. Start slow and increase the pace.
4. While verbalizing the directions, use one or more body parts to point in the direction of the arrows or triangles. Use a metronome to keep pace.
5. Rather than using a left to right and top to bottom progression on the chart, choose another orientation and perform the above activities. For example, have the child read the chart in columns from right to left and from bottom to top.

6. Have the child tell you the number of fingers raised on the hands while pursuing a left to right and top to bottom sequence on the chart.
7. Alternately raise the right and left hands with the proper number of fingers raised as you vocalize the number of raised fingers. Use a metronome to maintain consistent rhythm.

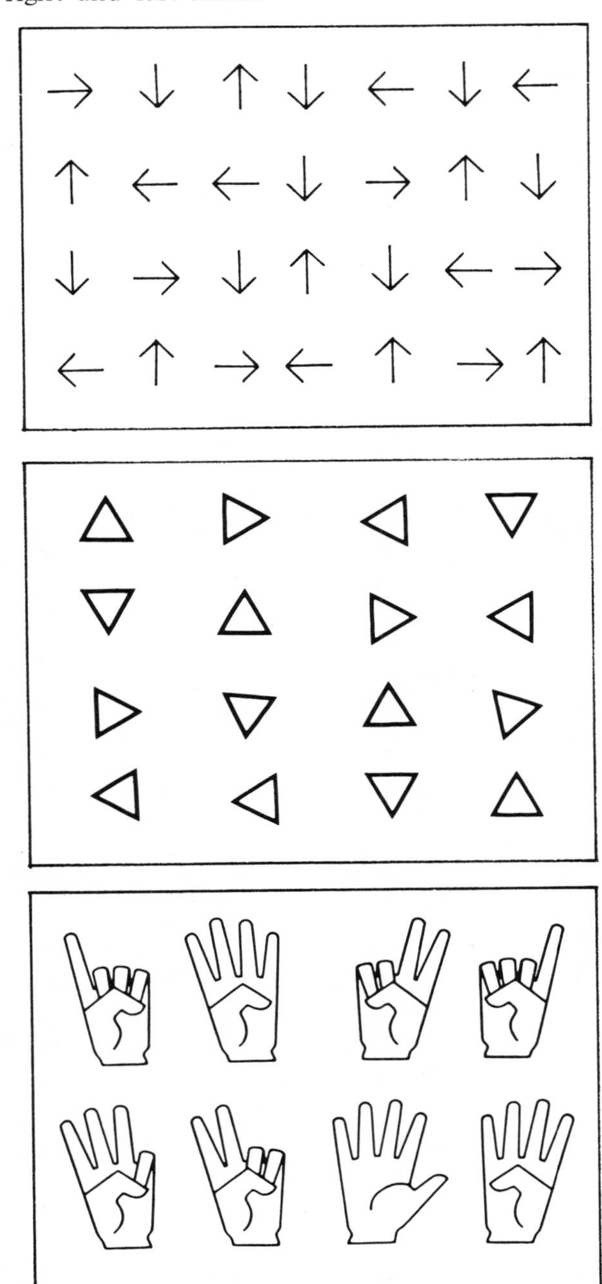

FIGURE 4.1. **Examples of charts or transparencies containing arrows, triangles, and fingers.**

PARQUETRY BLOCKS

CONSTRUCTION:

Parquetry blocks can be purchased commercially. However, they can be made with wood scraps. Size is variable, but a 16 inch tray is recommended as a container. A design can be painted on the tray as a guideline for the child to replace the blocks. Use the picture as a visual reference. The blocks to fit in the tray should be squares, triangles, and diamonds. They should be painted in various colors to aid in the discrimination of the shapes. Each shape is uniform in size and depth. The squares are 2 inches. The equilateral sides of the triangles are 2 inches. Each side of the diamonds is 2 inches. See Figure 4.2.

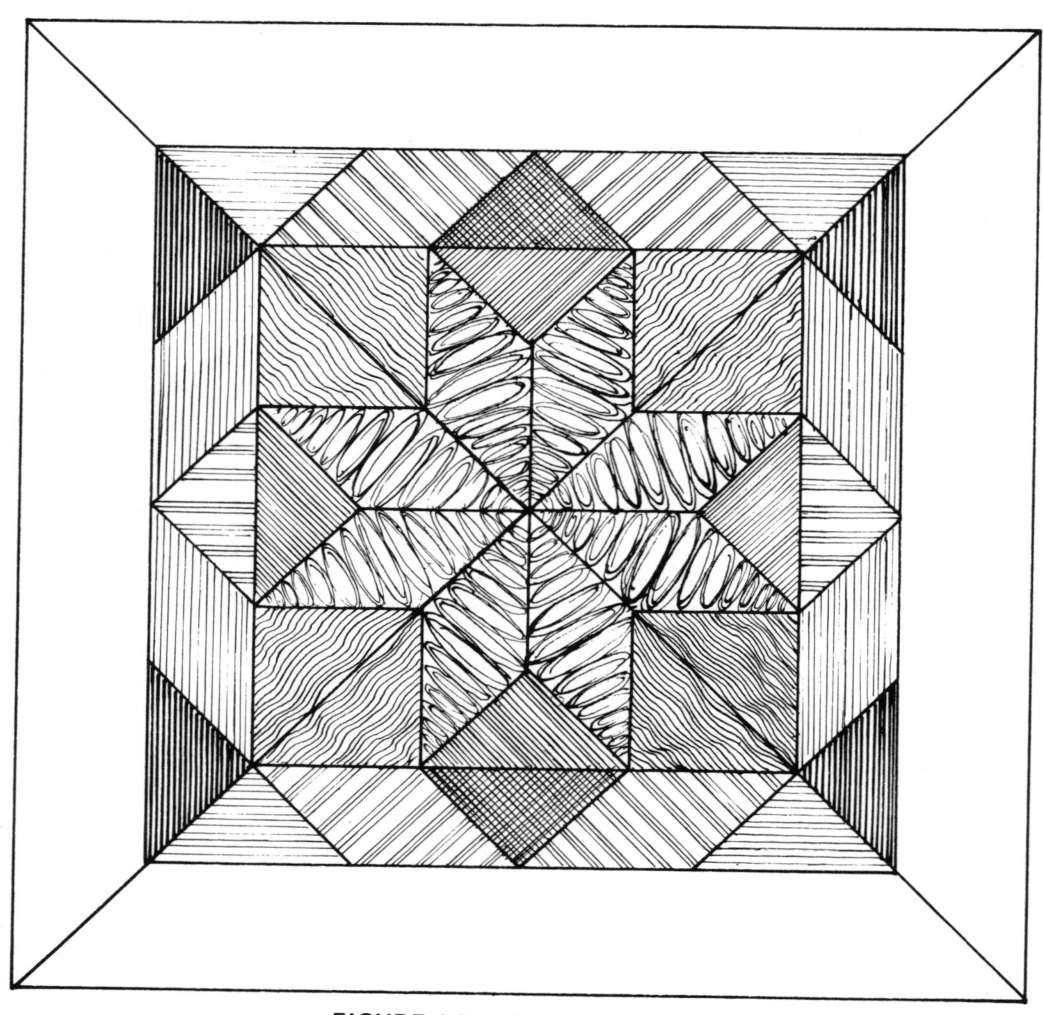

FIGURE 4.2. Paraguetry blocks

OBJECTIVES:

1. To develop spatial awareness concepts.
2. To increase a child's awareness of form perception.
3. To increase a child's awareness of tactile-kinesthetic perception.
4. To develop fine-muscle coordination.
5. To develop visual and auditory sequential memory.

ACTIVITIES:

1. Dump the blocks out of the tray. Have the child put them back in the tray using the design as a visual cue. Additional designs or patterns can be made on construction paper and inserted in the tray. Have the child devise a new pattern by fitting the pieces in the tray with no design to follow.
2. Put the blocks in a pile. Have the child give you all of the triangles, diamonds, or squares.
3. Have the child close his eyes and find a square, triangle, or diamond. Increase the difficulty by giving the child several requests. Have him find the forms and give them to you in the right order. For example, "Give me two squares, one triangle, and three diamonds."
4. Make a design with three or four blocks placed on a table. Have the child look at the design and make one exactly like it in shape, color, and organization, to the side of it.
5. Set up a two, three, or four piece pattern with the blocks. Have the child look at the pattern, close his eyes, and reproduce it by using visual sequential memory and tactile-kinesthetic sense.
6. Give auditory instructions indicating, color, shape, and placement for a sequence of two, three, or four blocks to be made into a design. Have the child reproduce the design from auditory sequential memory.
7. Perform the same previous activity, only use written task card instructions for visual sequential memory.

BLOCKS

CONSTRUCTION:

The blocks can be made or bought commercially. Obtain wood scraps from the local lumber yard, cut them to the desired shapes as seen in the pictures, and put several coats of varnish on them. Some of the blocks should have holes drilled through them on different sides. Narrow wooden rods or dowels, slightly smaller than the holes in the blocks, should be cut in varying lengths for additional building possibilities. See Figure 4.3.

Various shapes without holes:

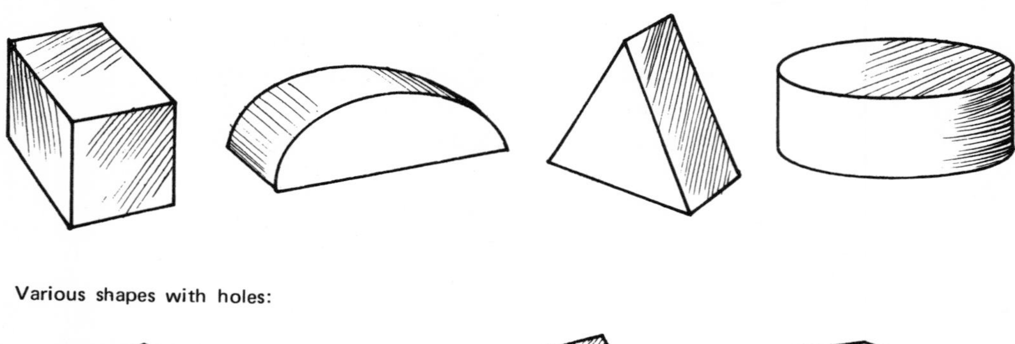

Various shapes with holes:

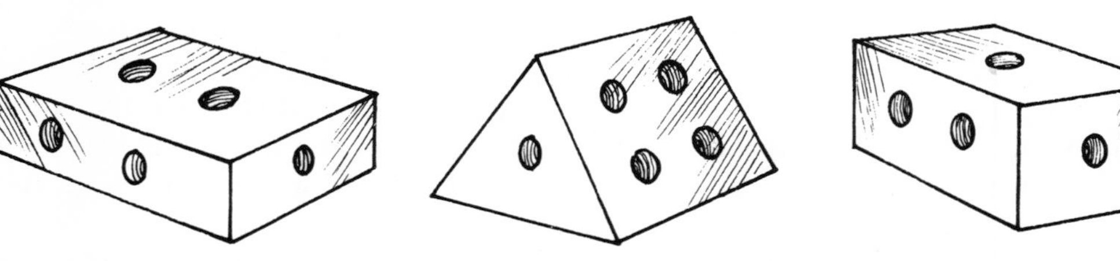

Various shaped blocks with pegs:

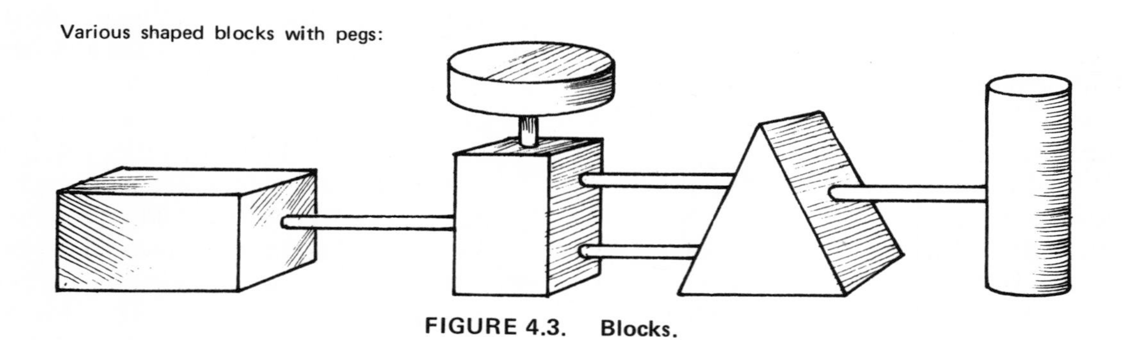

FIGURE 4.3. Blocks.

OBJECTIVES:

1. To develop spatial relationships concepts.
2. To develop laterality and directionality concepts.
3. To develop form perception.
4. To develop fine-motor and eye-hand coordination.
5. To increase tactile-kinesthetic awareness.

ACTIVITIES:

1. Have the child construct a design with the blocks and have him/her tell a story about it.
2. Use any forms desired and stack the blocks as high as possible. Combine terms, such as high and narrow or low and wide, as you challenge the child to stack the blocks.

3. Have the child copy a design made by you, another child, or from a picture.
4. Place several shapes in front of the child. Have him close his eyes and then ask him to hand you different shapes.
5. Begin a design and have the child complete it.
6. If the blocks vary in size and/or color, have the child classify the blocks according to color, shape, weight, and so on.

NAILBOARDS

CONSTRUCTION:

Two boards, approximately 16" x 16" x 3/4", wood nails, and a box of rubberbands are needed. The boards should be sanded and painted. Hammer the nails into the board, 1 inch apart in 12 rows of 12 nails. If you want to make the task simpler for the child, space the nails 2 inches apart or use fewer nails. See Figure 4.4.

OBJECTIVES:

1. To develop an awareness of spatial relationships.

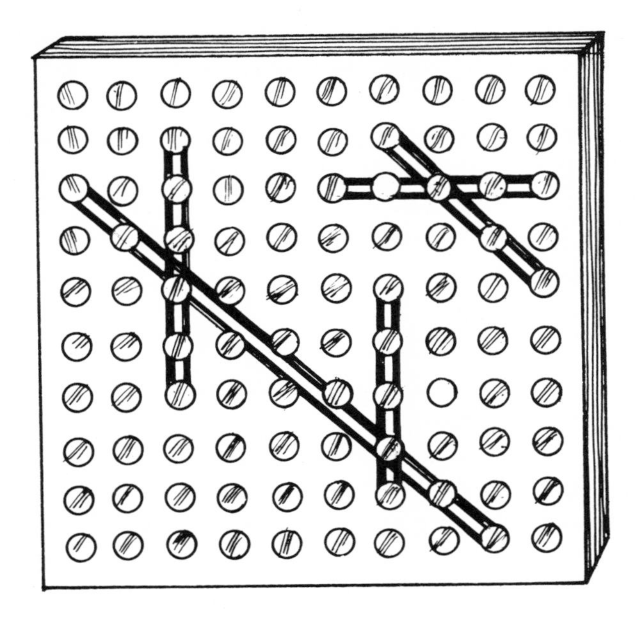

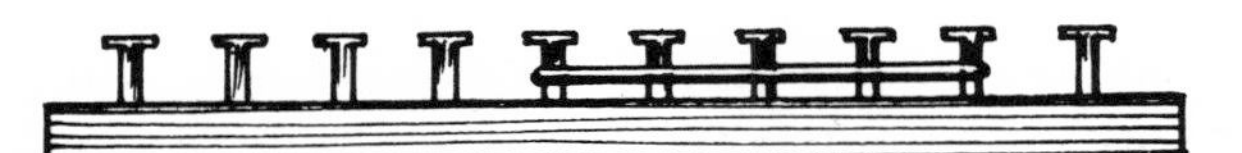

FIGURE 4.4. Nailboards.

2. To develop form perception.
3. To increase a child's fine-motor coordination.
4. To develop visual memory and sequencing skills.
5. To develop figure-ground perception.

ACTIVITIES:

1. Make a pattern on the nailboard using the rubberbands. Give the child another board and have him copy your design.
2. Give oral instructions concerning a design and have the child reproduce it.
3. Begin a geometric form and ask the child to complete the shape.
4. Stretch the rubberbands over the nails to make letters, numbers, or simple pictures, such as houses.

* Refer to geo-boards and golf tee boards for additional suggestions. (See pp. 52-53 and 71-72.)

GEO-BOARDS

CONSTRUCTION:

A pegboard, approximately 16 x 16", 25 nuts and bolts, rubber bands, and a notebook are needed. The task can vary in difficulty and can be regulated by the number of nuts and bolts screwed onto the board (four to twenty-five). Put a nut through a hole from the bottom of the board. Screw the bolt down flush to the board. Complete the board by inserting the remainder of the desired nuts and bolts to make a geo-board with 4, 9, 16, or 25 pegs. A notebook should be developed with a series of pictures, beginning with the simplest (4 dots) to the most complex pattern (25 dots), showing patterns to be duplicated with rubberbands on the geo-board. See Figure 4.5.

OBJECTIVES:

1. To increase the child's awareness of spatial relationships.
2. To develop form perception.
3. To develop fine-motor coordination.
4. To develop visual memory and sequencing skills.
5. To develop figure-ground perception.

ACTIVITIES:

1. Have the child use rubberbands to form the picture that he sees in the notebook. As the child improves, increase the level of difficulty by adding more nuts and bolts to the geo-board.
2. Show the child a picture and have him study it. Remove the picture and have him attempt to duplicate it from visual memory.

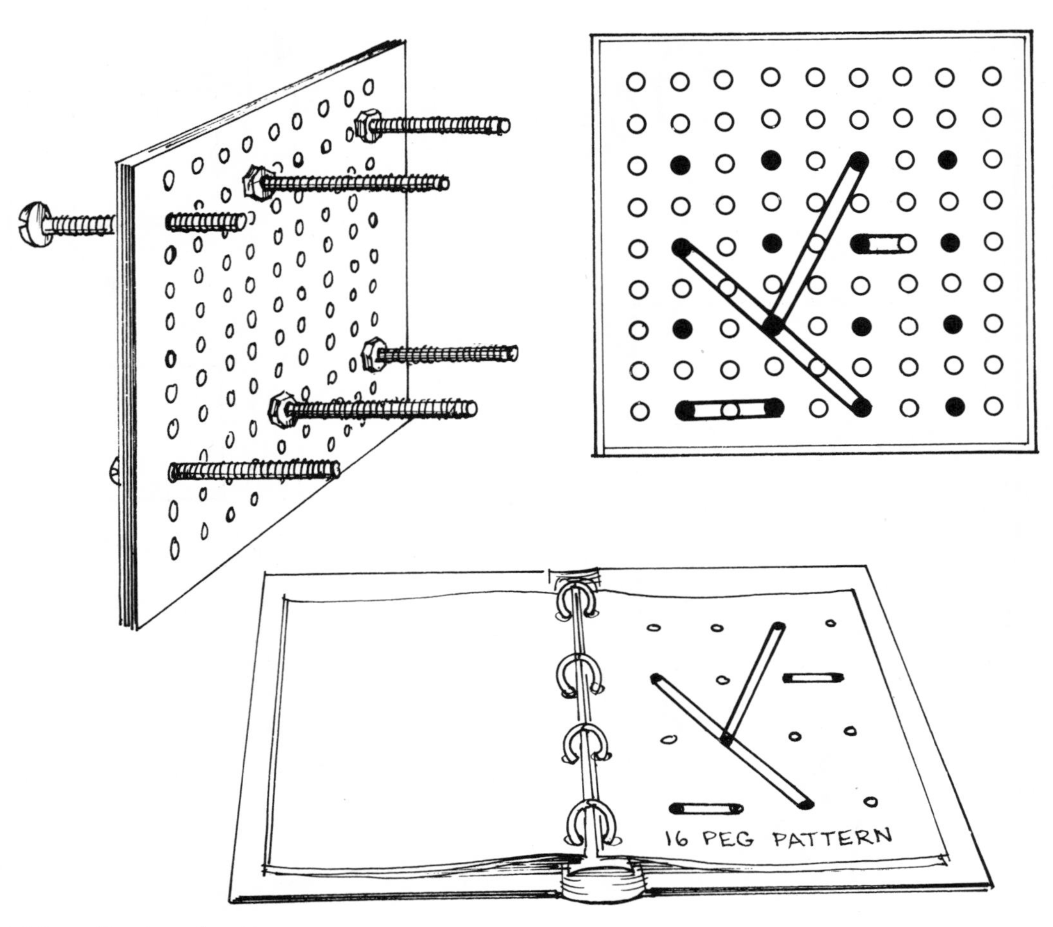

FIGURE 4.5. Geo-boards.

3. Have the child close his eyes. Give him oral directions about where to place the rubberbands. Then, change roles with the child and have him tell you where to put the bands. Compare the results with the actual picture to determine the correctness of the responses on the geo-board and the exactness of the directions.
4. Instruct the child to make a square, triangle, diamond, and the like on the geo-board by stretching the rubberbands in various ways.
5. Make a design using several rubberbands. Hand the board to the child and see if he can identify all of the forms.

* Refer to golf tee boards and nailboards for further suggestions concerning activities. (pp. 71-72 and 51-52.)

MAPPING EXPERIENCES

CONSTRUCTION:

Use rubber matting, carpet samples, or a similar material and cut it into the shapes of simple geometric forms or feet. A nonskid material will work best. Place

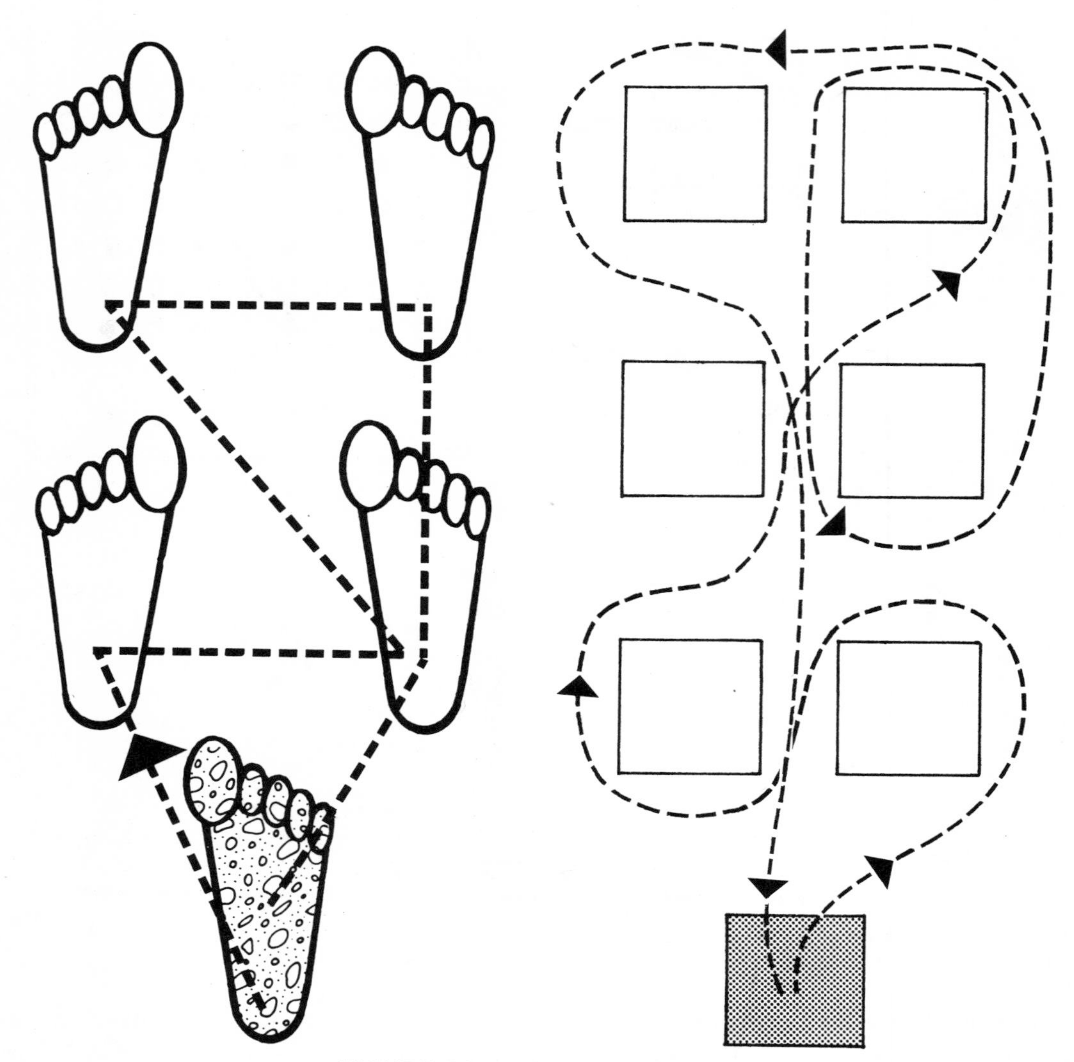

FIGURE 4.6. Mapping experiences.

the forms on the floor in a pattern, as indicated on index cards already designed. Draw map guidelines on 3" x 5" index cards, as seen in the picture. Homebase should always be a different shape or color to serve as a reference point for the child. Some of the map guides should direct the child to step on each of the forms. Others should direct the child to go around the forms in different directions, such as making inside or outside loops. Make up stories about farming, driving trucks, flying an airplane, and so on to accompany the map experiences. See Figure 4.6.

OBJECTIVES:

1. To develop spatial awareness concepts, such as laterality and directionality.

2. To develop mapping and orientation skills.
3. To help a child develop the ability to memorize patterns and sequences.
4. To enhance a child's awareness of figure-ground perception.

ACTIVITIES:

1. Lay three to ten squares on the floor. Hand the child an index card or map while he is standing at homebase. Tell him to follow the sequential guidelines from beginning to end. The task can be regulated according to the child's ability by increasing or decreasing the number of forms or guidelines.
2. Show the child a map. Then, take away the map and have him attempt the pathway from visual memory.
3. Have the child fabricate a story about his map experiences. If he intends to farm, he could plow various fields. If he plans to be a truck driver, he could make deliveries. Have the child think of other situations relating to the mapping experiences.
4. Have the child remove his shoes and follow the path with his eyes closed.
5. Pretend the forms are stepping-stones across a creek and have the child hop from one form to another.
6. Give the child a task card with written instructions instead of pictorial directions.
7. Use auditory instructions to guide the child through the map.

TWIST

CONSTRUCTION:

Purchase a drop cloth or a large sheet of plastic about 5 to 6 feet square. Use contact paper that has an adhesive backing to cut out triangles, circles, and squares of different colors. Attach them to the plastic, approximately 12 inches apart. The edges of the plastic sheet should be taped for durability. See Figure 4.7.

OBJECTIVES:

1. To develop spatial awareness, laterality, and directionality.
2. To increase form perception.
3. To develop gross-motor coordination and balance.
4. To develop body image and awareness.
5. To develop auditory memory and sequencing skills.

FIGURE 4.7. Twist.

ACTIVITIES:

1. One to four children can participate. More than one child makes it more exciting. Using a cardboard wheel, the teacher should spin the arrow and give directions to the children as to which hand or foot to place down. The child is to put his body part on the designated spot and keep it there. The winner is the one who does not lose his balance and fall while reaching for a new spot. Often, it is more exciting if the teacher just makes up new directions. This way he can watch the children and make

the moves more difficult and help to tangle the children even more.

2. Play a modified game of hopscotch by hopping barefoot on the forms.

SIMILARITIES AND DIFFERENCES PUZZLES

CONSTRUCTION:

Sturdy cardboard, magic markers, and construction paper are all that is needed to develop these puzzles. Puzzles can be from three to six pieces and cut in the shape of squares, circles, rectangles, and the like. Cut out the desired cardboard size and shape. Then, let your creativity take over. Draw pictures with the magic markers on the cardboard or cut out forms with the construction paper or cartoon characters from the comic strips and coloring books and paste them on the cardboard backing. Make all of the pictures from the puzzle the same, except one, as seen in Figure 4-8 p. 58. Use a code on the back for reinforcement in telling the child when he has the correct answer.

OBJECTIVES:

1. To enhance the development of laterality and directionality concepts.
2. To increase the child's awareness of position in space.
3. To increase a child's awareness of form perception and figure-ground perception.

ACTIVITIES:

1. Give the child one of the puzzle sets. Have him lay the pieces out in front of him. Then, have him select the only different puzzle piece.
2. Have the child draw a picture similar to the puzzle pieces.

FOLLOWING DIRECTIONS

CONSTRUCTION:

Develop a series of maps similar to the ones in the pictures to help the child learn the concepts of left, right, north, south, east, west, up, down, and so on. Put the maps into a notebook. Use a piece of clear plastic acetate over the notebook sheets so that they can be used again. Have the child use a crayon, grease pencil, or magic marker on the plastic acetate so that the marks can be erased by rubbing them with a piece of cloth.

FIGURE 4.8. Similarities and differences puzzles.

OBJECTIVES:

1. To increase a child's directional awareness.
2. To develop visual and auditory sequential memory.
3. To develop fine-motor coordination.
4. To help children develop auditory encoding abilities.

ACTIVITIES:

1. Have the child listen to verbal instructions and perform the prescribed tasks on the pictures. Speak to the child or use tape recorded instructions.
2. Give the child a task card. Have him read and follow the directions on the card.
3. Increase the task's difficulty by removing the directional words on the map. The child must now develop an awareness of direction without using the word as a cue.

Ice Is Fun (See Figure 4.9)

   a. At the upper left hand side of the lake, Snoopy ran onto the ice. Draw a line from where Snoopy ran onto the ice to the upper right hand side of the lake.
   b. Snoopy then changed directions and slid across the lake moving from right to left. Draw a line showing Snoopy's direction.
   c. Snoopy changed directions again and slid across the lake, moving from left to right. Draw another line showing his movement.
   d. Snoopy then ran towards the lower part of the lake. When he got to the middle, he fell on his back. Draw a line showing Snoopy's movement.
   e. After picking himself up, Snoopy ran off the lake at the lower left hand side and fell on his face. Draw a line to show where Snoopy landed.

Following Directions (See Figure 4.10)

   a. In the first row, color in the top half of the third square from the left.
   b. In the third row, color in the right half of the fourth circle.
   c. In the sixth row, find the second square from the left. Draw a diagonal line in the square, from the upper right corner to the lower left corner.

4. Give the child a sequence of instructions and see if he can follow them in the correct order.
5. Have the child dictate a set of directions to another child or you, as the teacher, to help the child develop auditory encoding concepts.

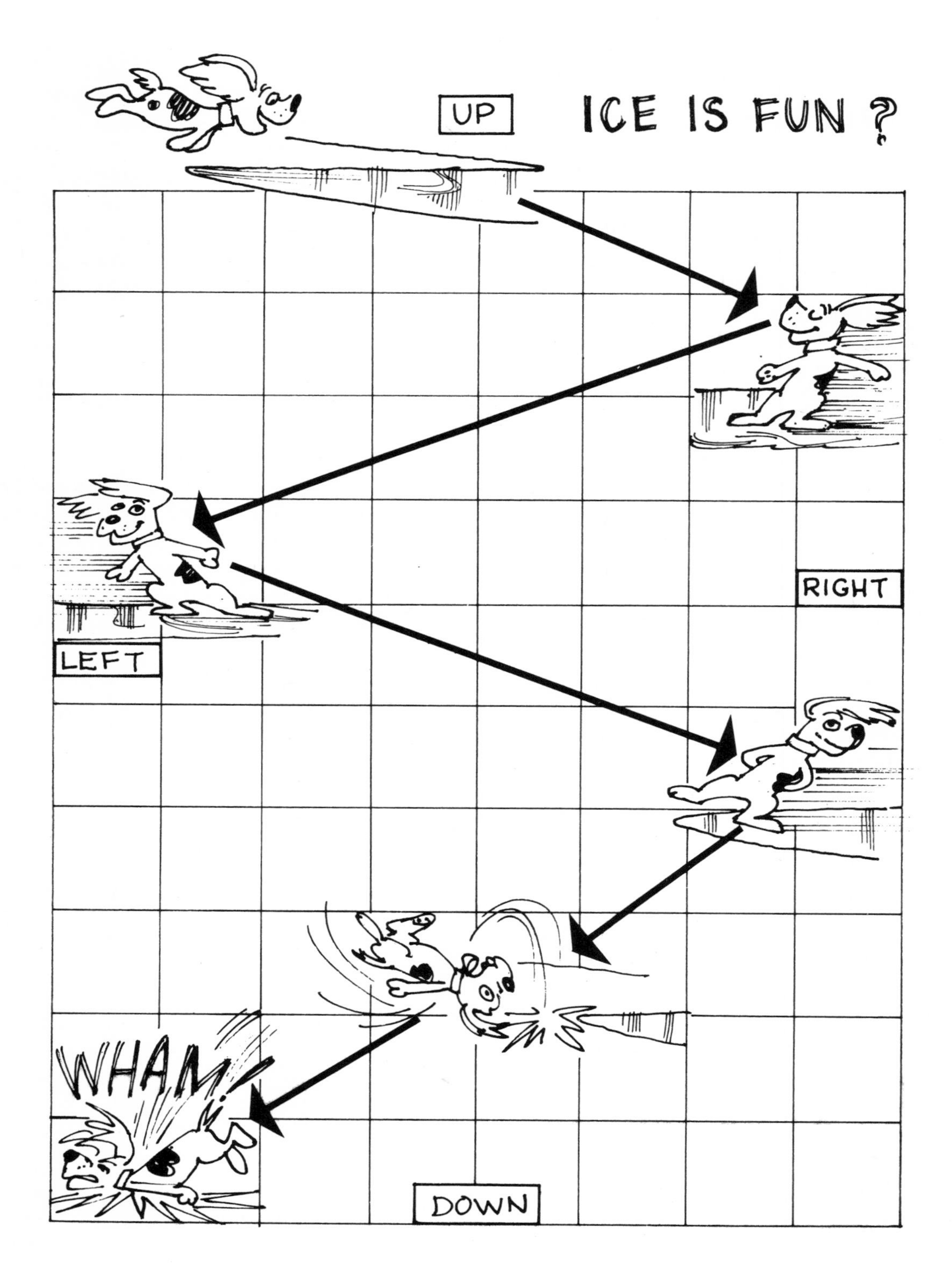

FIGURE 4.9. Ice is fun?

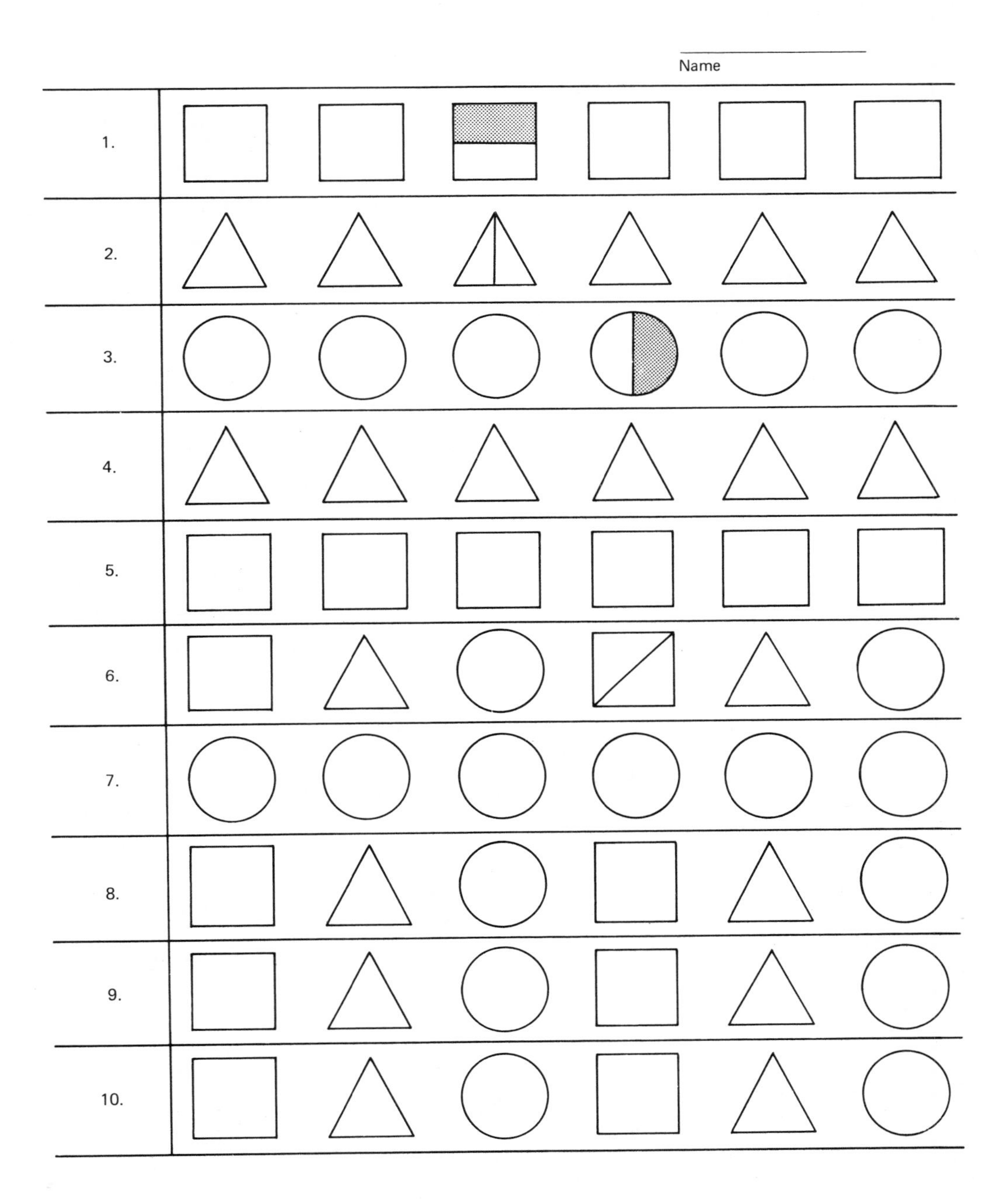

FIGURE 4.10. Following directions.

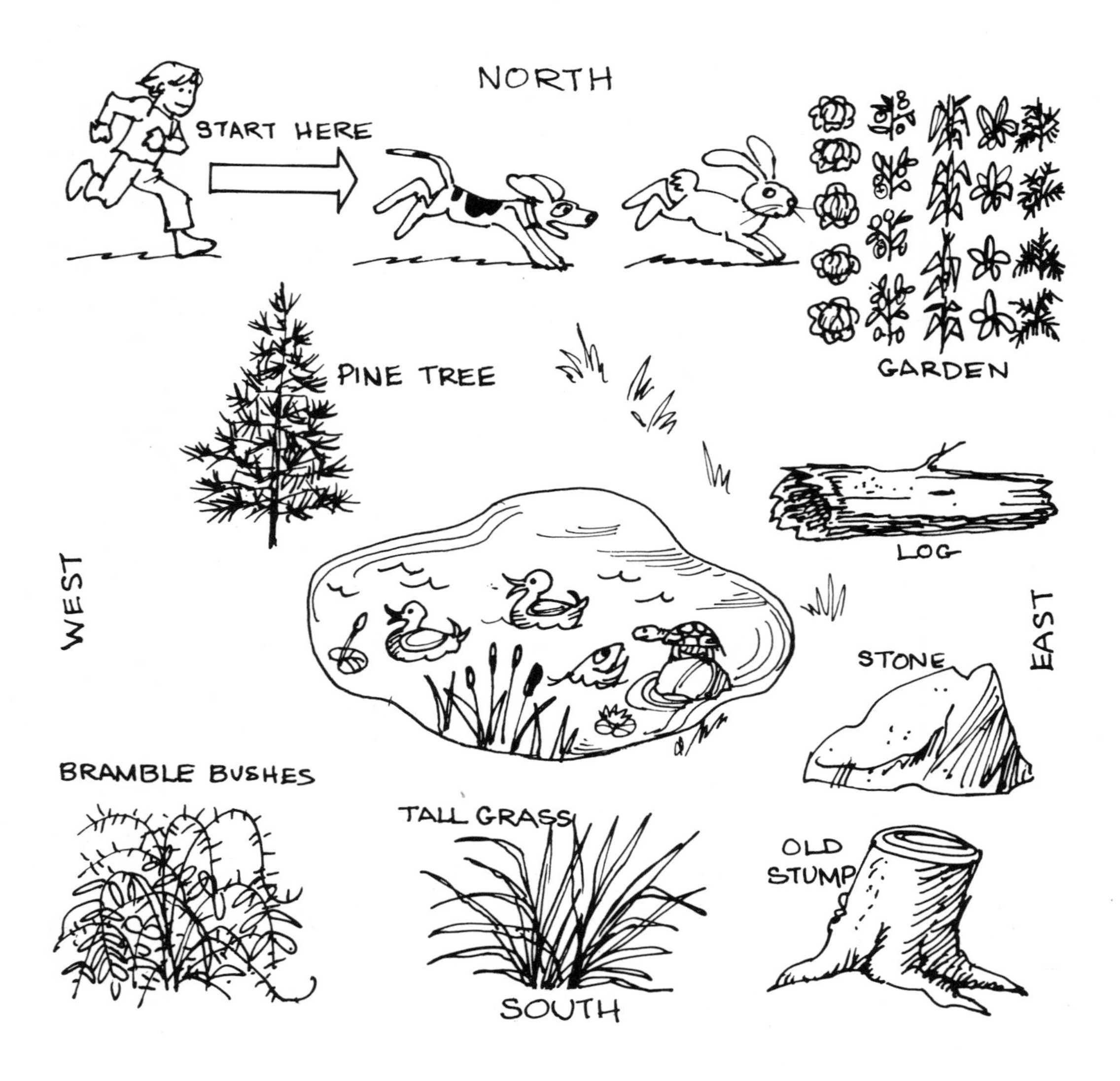

FIGURE 4.11. Finding directions. Written upside-down. Face north. South is behind you; east is to your right; west is to your left. Look at the map of the field below. The directions *north, south, east,* and *west are marked on the map.* Use these directions to find Peter's dog.

Once, Peter's dog chased a rabbit. Peter ran after his dog. Draw a line to show where Peter's dog and the rabbit went. The directions in the sentences tell you where to draw your lines. a: The dog chased the rabbit *east* into the garden; b: the rabbit hopped *south* to the log; c: The dog chased the rabbit *west* to the pine tree; d: The rabbit ran *south* to the bushes; e: The dog followed the rabbit *east* to the tall grass; f: The rabbit hopped *north* to the pond; g: The dog chased the rabbit *east* to the stone; h: The rabbit ran *south* to the stump and popped down a hole; i: The dog ran to the stump, then *west* to the tall grass to meet Peter.

EUCLIDEAN GEOMETRY GAMES

CONSTRUCTION:

A chalkboard and chalk or a paper and pencil are the only requirements for this experience.

OBJECTIVES:

1. To help a child develop concepts of position in space and spatial relationships.
2. To increase a child's awareness of form perception.
3. To refine visual perception concepts in the child.

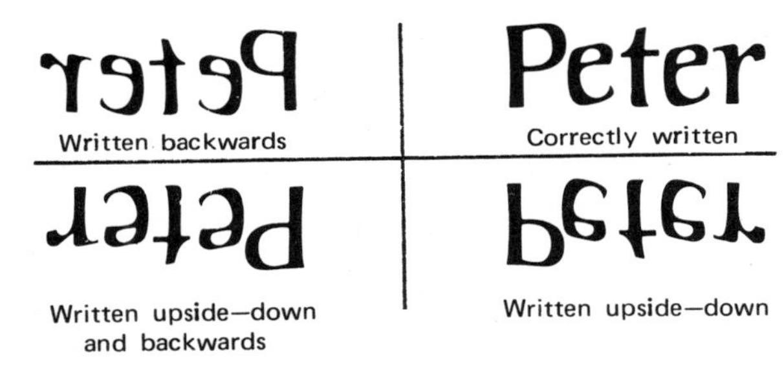

FIGURE 4.12. Euclideon Geometry Games.

ACTIVITIES:

1. Draw two perpendicular lines that bisect one another on the board or paper. In the first quadrant, have the child draw his name, as seen in Figure 4.12. In the second quadrant, have the child draw his name backwards. On the third quadrant, his name should be upside down and backwards and a mirror image of the second quadrant. On the fourth quadrant, his name should be drawn upside down—a mirror image of the first quadrant. See Figure 4.12.
2. Perform the same activities with spelling words, numbers, and simple designs. Move from simple to complex activities until the child is familiar with all rotational aspects of space. Let the child use a mirror to check his work.

5

Form Perception

Form perception is the ability to recognize and reproduce geometric forms of varying complexities. From reviewing the literature, it appears that children go through three stages in the development of each of these concepts. As children learn to recognize objects, it is thought that the first stage is "globular," as the small child views ill-defined masses or objects. This stage is similar to that of the small infant who learns to interpret those sources of nutrition and care in the environment as mother. Gradually, persons in the environment become more defined, as infants become capable or recognizing more details. Individuals are differentiated and strangers even frighten babies of several months.

Objects and geometric shapes also become differentiated during this second stage of form recognition called syncretic form. At this stage, objects are recognized as wholes being the sum of parts or elements. This stage is limiting, however, because if one element of the object is changed the child may no longer recognize it. For example, it may be shocking for the child the first time he sees mom wearing a wig or dad without glasses. The child may also fail to recognize a geometric shape if the size or color is changed. Once the child reaches the third stage of form recognition (constructive form), he can recognize a particular object regardless of the color, size, texture, and so on. The child can even recognize an object if part of it is hidden from view. At this stage, the whole becomes greater than the sum or it's parts (27).

The ability to reproduce forms also follows three different stages. The first is the ability of the child to trace around or over objects used as guidelines. This implies that there are some physical boundaries for the child to use, such as a template. The second level of form reproduction is copying. At this stage, the child is allowed to view a particular pattern and then copy one off to the side while using proper color, size, positioning, and the like. At the third stage of form reproduction the child is able to produce forms from memory. Thus you may ask a child to reproduce a picture for you; or you may show him a picture, remove it from sight, and require the reproduction of the picture from visual memory (27).

Children learning to reproduce forms progress from a scribbling stage to a level where he/she is able to draw symmetrical shapes (11, 12, 27). The curved lines are gradually enclosed to form circles. The horizontal and vertical lines are combined to form crosses, squares, rectangles, and the like. Diagonal lines are the last to be mastered, and as a result, figures like a triangle and diamond are the last to be mastered. Once the basic figures can be drawn, the child can learn to combine forms and reproduce letters of the alphabet, numbers, and pictures. In the case of drawing pictures, children are able to pay more attention to refined details as they master the elements. For example, the child may move from drawing a stick person to adding features, such as hair, a face, a trunk, hands, feet, clothes, and the like. When drawing other pictures, such as toys, animals, or houses, the figures also progress from stick drawings to more detailed forms.

OBSTACLE COURSE FORMS

CONSTRUCTION:

Use pieces of cardboard from large appliance boxes or thin sheets of styrofoam to construct the obstacle course forms. Circles, triangles, squares, diamonds, rectangles, and other forms can be cut out. Vary the dimensions from 2 to 4 feet. Allowing for a 2 inch border and cut out the center of the form so that the child can crawl through it. A base should be made for each form from a piece of 2" x 4", as seen in the picture. Each base should be about 4 inches long. A groove should be cut in each base, ½ inch deep and ¼ inch wide. Insert the forms into the bases and make an obstacle course. See Figure 5.1.

OBJECTIVES:

1. To increase a child's awareness of form perception.
2. To increase a child's spatial awareness.

FIGURE 5.1. Obstacle course forms.

3. To enhance a child's body awareness or self-concept.
4. To develop gross-motor coordination and agility.

ACTIVITIES:

1. Tell the child to crawl feet first or head first through each of the forms. Challenge him to change body positions and have his front or back facing the floor as he passes through the forms. Tell the child to try not to touch the forms as he passes through them.
2. Set up an obstacle course using the forms and other objects, such as balance beams, spring boards, chairs, and the like. As the child's skill improves, time him to see how long it takes him to progress through the course.
3. Lay the forms on the ground and use them for self-testing activities. For example:
   a. Stand in the center of the form. Can you balance on one base of support? Can you bounce a ball while standing or kneeling within your form?
   b. Can you jump in and out of the form without touching it? How many ways can you move around the form? How many ways can you move in and out of the form on two bases of support?
4. Lay the forms on the ground and use them to play hopscotch.
5. Use the forms as templates and perform the suggested activities.
6. Use the forms as targets and try to throw beanbags through them at varying distances.

MAILBOXES

CONSTRUCTION:

Use shoe boxes, soap boxes, or other small cartons to make this piece of equipment. Cut a slit in one of the sides of the box, as seen in Figure 5.2. The side of the box opposite the slit should be removed for easy access. After the "mailboxes" are made, make cards or "mail" that will fit through the slits in the mailboxes out of construction paper. The cards should be made in a series and should consist of forms, letters, or numbers that children are learning to recognize. Similar designs, such as p, b, d, q, or 6 and 9, should be grouped together in a series. Other series might be circles and ovals, squares and rectangles, diamonds with different orientations, m and w, n and z, L and V, 1 and 7, and 2 and 5. About six cards of each form, letter, or number in each series should be made. Each card should have a code placed on the back. Each card should have a corresponding box with the same code drawn on the bottom inside, as seen in the figure. Use a paper clip or glue to fasten one of the matching cards to the parent mailbox. See Figure 5.2.

OBJECTIVES:

1. To develop form perception.
2. To increase a child's awareness of laterality and directionality.
3. To develop a child's awareness of position in space and spatial relationships.
4. To increase a child's eye-hand and fine-motor coordination.

ACTIVITES:

1. Select a series of common designs, such as p, b, d, and q. Be certain the proper designs correspond with the correct mailboxes so that the codes will match when the child makes a correct response. Give the child the remainder of the cards in the sequence and have him try to place each card in the appropriate box. The child can check his response with the code after each attempt to check his accuracy. Two children can also participate at the same time. One child can be the mailman and put the mail in the boxes. The second child can be the postmaster and check the codes to see if the mail was placed in the right box.
2. Perform the same activities with other letter series, number series, and form series.

TEMPLATES

CONSTRUCTION:

Pieces of plywood, masonite, redwood, pine, wall board, and the like, varying between 1/8 and 1 inch in thickness are needed to make templates. The

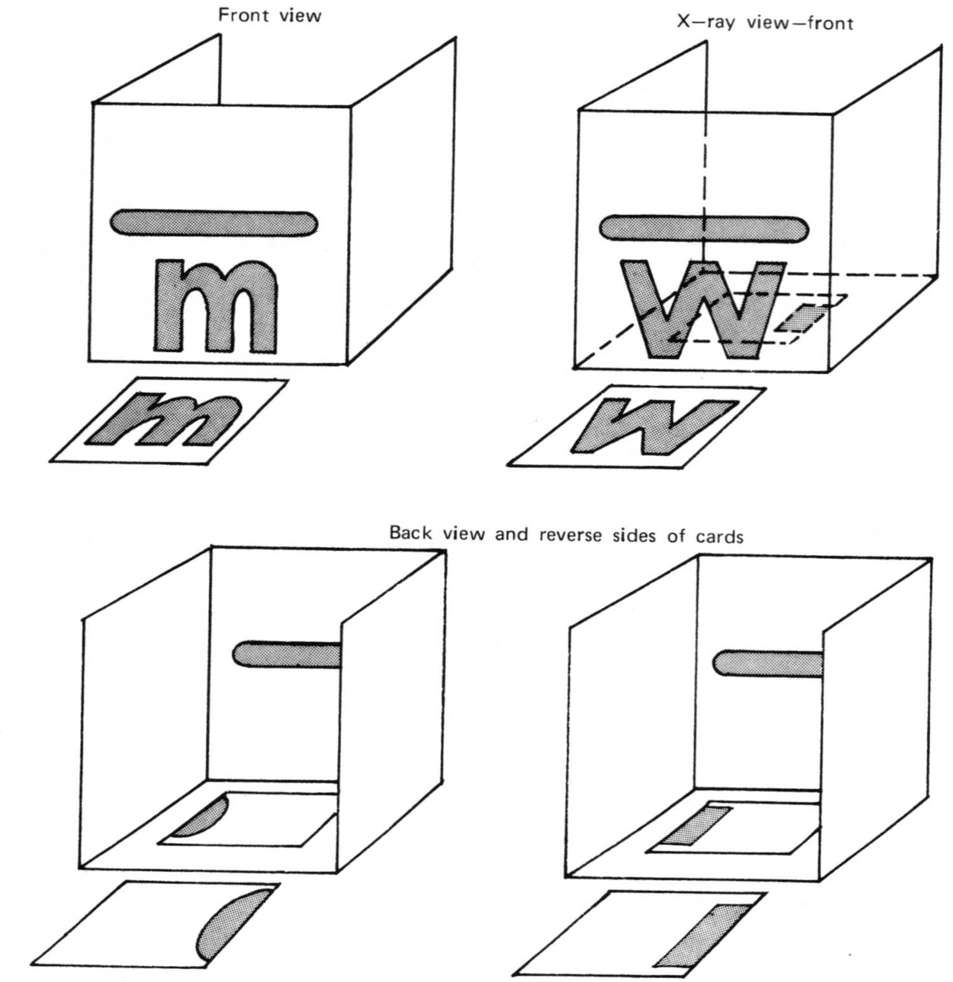

FIGURE 5.2. Mailboxes.

dimensions of the templates vary according to the size of the wood, but are usually from 6 to 18 inches in diameter. Cut the template form (circle, square, triangle, rectangle, or diamond) to the desired size. About 1½ to 2 inches inside this shape, cut out an inner form of the same shape. Attach a handle to the inner form, as seen in Figure 5.3.

OBJECTIVES:

1. To aid in the development of form perception.
2. To develop fine-motor coordination.
3. To increase a child's awareness of tactile-kinesthetic sensations.

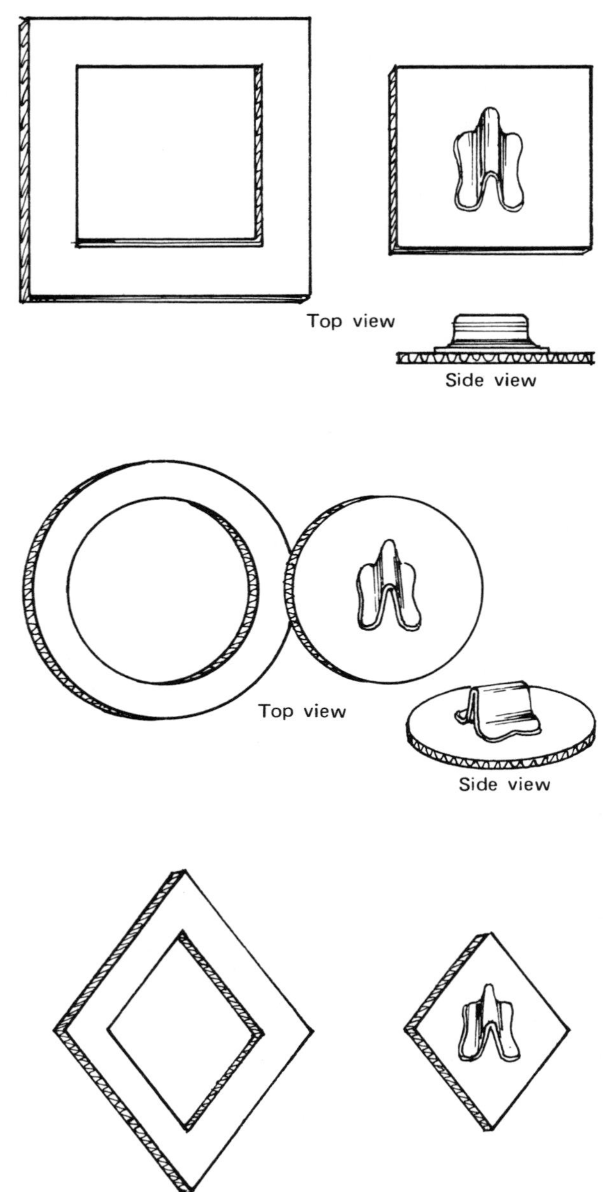

FIGURE 5.3. **Templates.**

ACTIVITIES:

1. Have the child pick up the outer form and trace around the inside or outside of the figure with his finger. Have him describe the characteristics of the figure to you. How does a circle feel? How many corners and sides to the triangles, squares, and rectangles have? Can you find other objects in your room that have the same shapes? Draw the forms in the air.
2. Trace around the outer edge of the inner form on the chalkboard or on your paper.
3. First hold the form with your nondominant hand and do the tracing with your preferred hand. Then, change hands.

4. Use different colors of chalk to trace around the inner and outer edges of the templates.
5. Once the image is made with the template, take the template away and trace over the image you have made. First use one hand; then, use the other. Change directions as you draw the figures.
6. Have the child close his eyes. Then give him a template. Can he tell you what form he was given? Have him describe it to you with his eyes closed. Have him draw the form on the board or paper while he has his eyes closed.
7. Show the child a template and ask him to draw an identical one for you on the board or paper. After he is finished, use the template to measure the results.
8. Have the child color in the area within the outer form on a sheet of paper. Then, have him cut out the form he made with a scissors.
9. Use the forms as targets and have the child throw beanbags or yarn balls through the holes in the outer forms at varying distances.

SAND, CLAY, AND FINGER PAINT

CONSTRUCTION:

There is no construction involved for the experiences in this project. It would be wise, however, to provide a sand-table or trays of sand to contain the needed materials. An art area for clay and finger paints is also recommended.

OBJECTIVES:

1. To develop form perception.
2. To increase fine-motor coordination.
3. To increase a child's tactile-kinesthetic awareness.
4. To develop a child's auditory and visual decoding abilities.
5. To develop a child's manual encoding abilities.

ACTIVITIES:

1. Give the child a sand tray. Show him a form, letter, or number and have him trace it in the air; then, have him trace it in the sand.
2. Give the child some clay and have him press it into the form, letter, or number desired.
3. Have the child draw the forms, letters, or numbers on the paper, using finger paint.
4. Have the child perform the latter experiences with his eyes closed.
5. Vary the size of the form, letter, or number. How small can the child make each figure in each medium? How large can he make them?

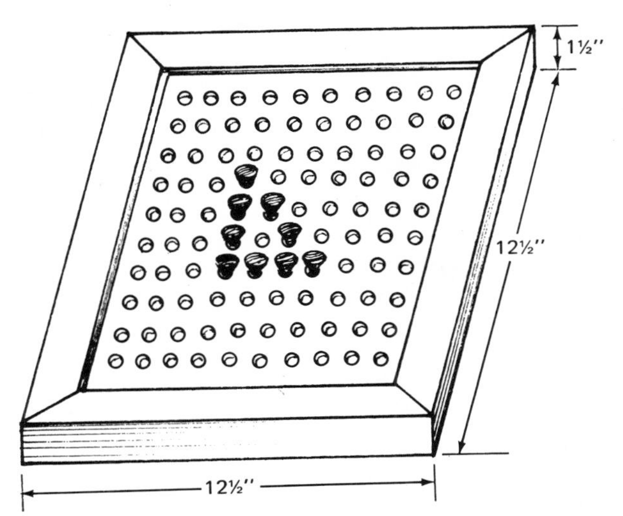

FIGURE 5.4. Golf tee pegboards.

GOLF TEE PEGBOARDS

CONSTRUCTION:

Purchase a piece of pegboard and a piece of wood 1½" x 1" x 10'. To make each pegboard, cut a 12 inch square. Cut the 1½ x 1" piece of wood at 45° angles to produce a frame. Glue or nail the pegboard to the frame. Purchase several dozen golf tees at a local sporting goods store. See Figure 5.4.

OBJECTIVES:

1. To develop form perception.
2. To increase fine-motor coordination.
3. To develop visual memory and closure.
4. To develop spatial awareness concepts involving laterality and directionality in children.

ACTIVITIES:

1. Instruct the child to place the pegs in a vertical and straight line on the pegboard. Change the instructions to form horizontal and diagonal lines.
2. Draw a form on a piece of paper. Place the form on the board and have the child punch holes through it to make the form.
3. Give the child verbal instructions on how to place the pegs to make a design.
4. Have the child close his eyes and make a form on the pegboard with the golf tees.
5. Have the child put the corners of a form on the pegboard. Tie a piece of yarn or string on one of the tees and wrap the line

around the corners to make the sides.
6. Begin a form for the child. Have him complete the form.
7. Give the child the bag of tees. Using both hands, see how many tees he can place on the board in ten seconds.
8. Make a picture or a design by placing the tees on the board.

* For further suggestions, refer to geoboards and nailboards in another section of this book (pp. 51-53.)

FORM, LETTER, AND NUMBER DOMINOES

CONSTRUCTION:

Homemade dominoes can be made from plywood, wall board, or any other type of wood that is 1/8 to 1/4 inches thick. Cut the wood into pieces approximately 2½" x 5". Drill small indentations for the numbers into the wood. on each half of the domino. Vary the numbers from one to six and leave an appropriate amount of blank pieces or zeroes. Paint the drilled indentations with bright colors so that the number of them on each half of the domino are easy to identify. Another way to make the number dominoes is to paint the numbers on each half of the pieces. For additional variations, form and letter dominoes can be made in the same fashion. Use a router to groove the forms or letters into the domino pieces. Then paint the grooves, or simply paint the forms or letters on the pieces of wood. In an effort to vary the texture, glue sandpaper, felt, or any other type of material in the shape of the forms, letters, or numbers to the domino pieces. See Figure 5.5.

OBJECTIVES:

1. To develop form perception.
2. To enhance figure-ground perception.
3. To increase eye-hand coordination.
4. To improve balancing skills.
5. To enhance concepts of laterality and directionality.
6. To increase the child's awareness of position in space and spatial relationships.

ACTIVITIES:

1. Play a game of dominoes by matching the ends of each piece. Use the numbers, number dots, forms, or letters.
2. Modify the game of dominoes by having the child find the mirror image of the form or dot pattern. Increase the difficulty by asking him to draw a dot or form pattern that is rotated 90° or 180° from the example. Check his results by rotating the domino to see if it matches with the picture he has drawn.
3. Build a structure or stack the dominoes as high as you can.

FIGURE 5.5. Form, letter, and number dominoes.

PUZZLES

CONSTRUCTION:

Many of the commercially made developmental puzzles for young children are very fine products. However, because there may be a lack of funds or because certain children have specific interests, equally useful puzzles can be made. The needed raw materials consist of corrugated cardboard, stiff poster board, glue, razor blades or cutting knives, and pictures. Select a picture that the child is interested in. Trucks, athletes, cartoon figures, scenery pictures, and the like, are often popular with children. Paste the picture to a stiff backing. With a sharp blade, cut out the puzzle pieces. Puzzles may consist of two simple and basic pieces or may be more complex with 15 to 50 pieces. Common objects, as puzzle pieces hidden among irregular shapes, help children to develop figure-ground perception. Laminate the puzzles with dry mount material or clear plastic for a more durable product. See Figure 5.6.

OBJECTIVES:

1. To enhance awareness of tactile-kinesthetic sensations.

FIGURE 5.6. Puzzles.

2. To increase form perception.
3. To develop visual sequencing abilities.
4. To enhance figure-ground perception.

ACTIVITIES:

1. Put the puzzle together.
2. Identify any hidden shapes within the puzzle (heart, dog, circle, and diamond).
3. Try to put a relatively easy, irregular shape puzzle together with your eyes closed.

ATTRIBUTE BLOCKS

CONSTRUCTION:

Use 1/8 and 1/4 inch press board to cut out the figures seen in Figure 5.7. Make sure to make two or more different sizes of the same shape and paint a selected

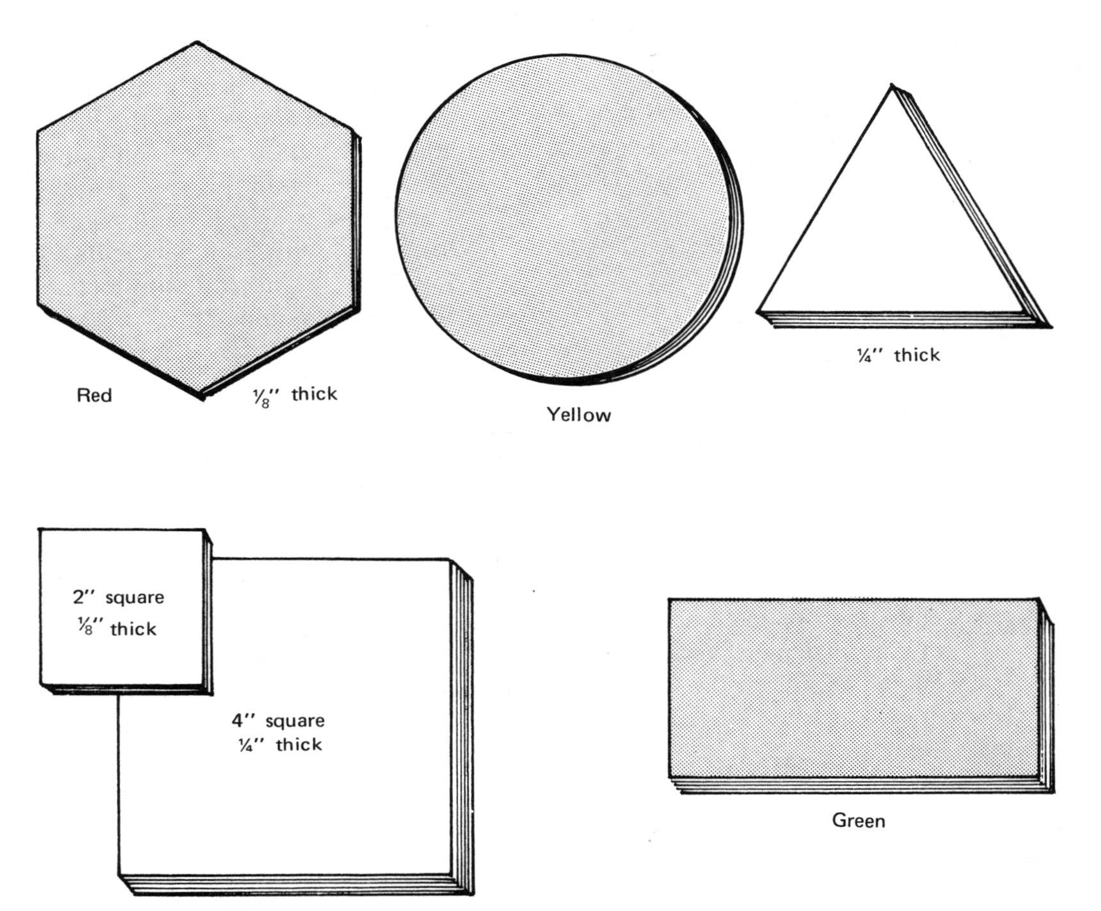

FIGURE 5.7. Attribute blocks.

number of each a different color. A full set of attribute blocks will contain at least 60 blocks—5 shapes x 2 sizes x 3 colors x 2 thicknesses.

OBJECTIVES:

1. To provide practice in classifying and coding information according to characteristics such as size, thickness, color, or shape.
2. To develop an awareness of form perception and form constancy.
3. To develop visual association and visual figure-ground perception.
4. To increase eye-hand and fine-motor coordination.

ACTIVITIES:

1. Have the child sort the blocks according to shape, size, or thickness. To help the child, the thick blocks may be coded by having the edges painted.
2. Have the child sort the blocks according to color.
3. Have the child sort the blocks into sets having two common properties. For example, the child could sort the blocks that are thick and yellow or thin and square.
4. Have the child play a game similar to dominoes, in which he develops a chain or sequence, with each succeeding block having at least one common characteristic of the preceding block.

TANGRAMS

CONSTRUCTION:

Tangrams are ancient Chinese puzzles consisting of five different basic geometric forms. These forms can be made from scrap pieces of plywood 1/4 to 3/8 inch thick. Cut each of the forms into the dimensions shown in Figure 5.8 and 5.9. For each set of tangrams, make four of the large triangles, two of the medium triangles, four of the small triangles, two squares, and two parallelograms. Paint one side of the forms with paint and the other side with varnish.

OBJECTIVES:

1. To help children develop better concepts of form perception.
2. To increase the children's awareness of position in space and spatial relationships.
3. To enhance development of visual and auditory sequential memory.

ACTIVITIES:

1. Assemble the puzzle pieces into various numbers, letters, and figures as seen in Figure 5.9.
2. Use auditory directions to give children clues as to how to construct a letter, number, or form.
3. Show the child a picture. Then, see if he can construct the puzzle from visual memory.

* For further suggestions, refer to parquetry blocks in another section of this book (pp 48-49).

ALPHABET BLOCKS

CONSTRUCTION:

Alphabet blocks can be made from scrap pieces of wood 1" thick. Cut each of the eight forms into the dimensions shown in Figure 5.10. Make six forms for

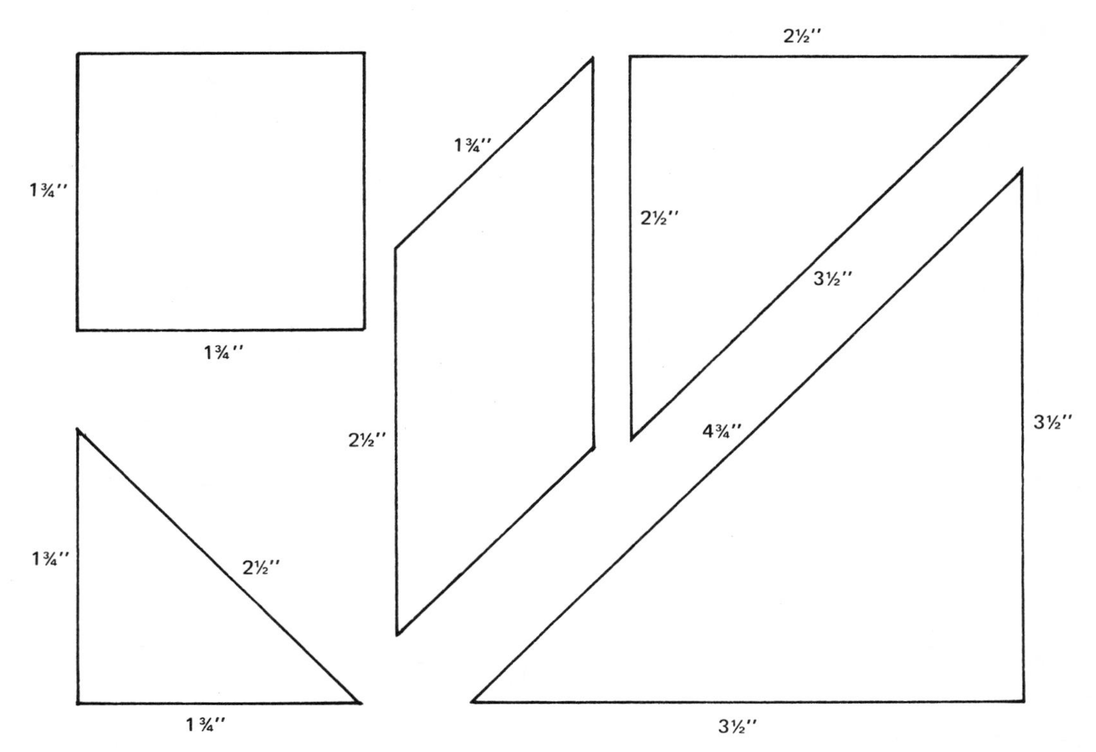

FIGURE 5.8. Pieces of tangram puzzles.

FIGURE 5.9. Tangram puzzles.

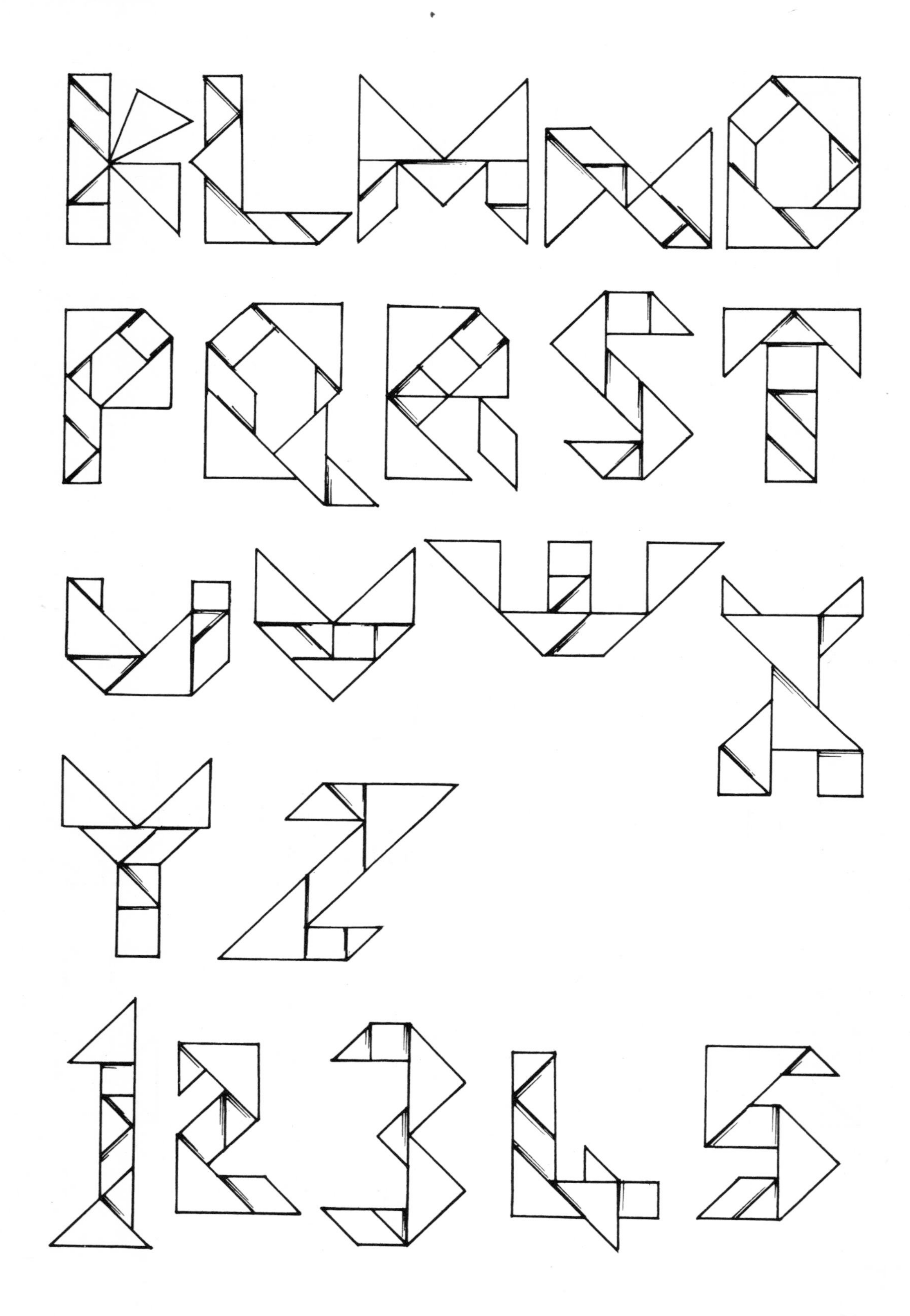

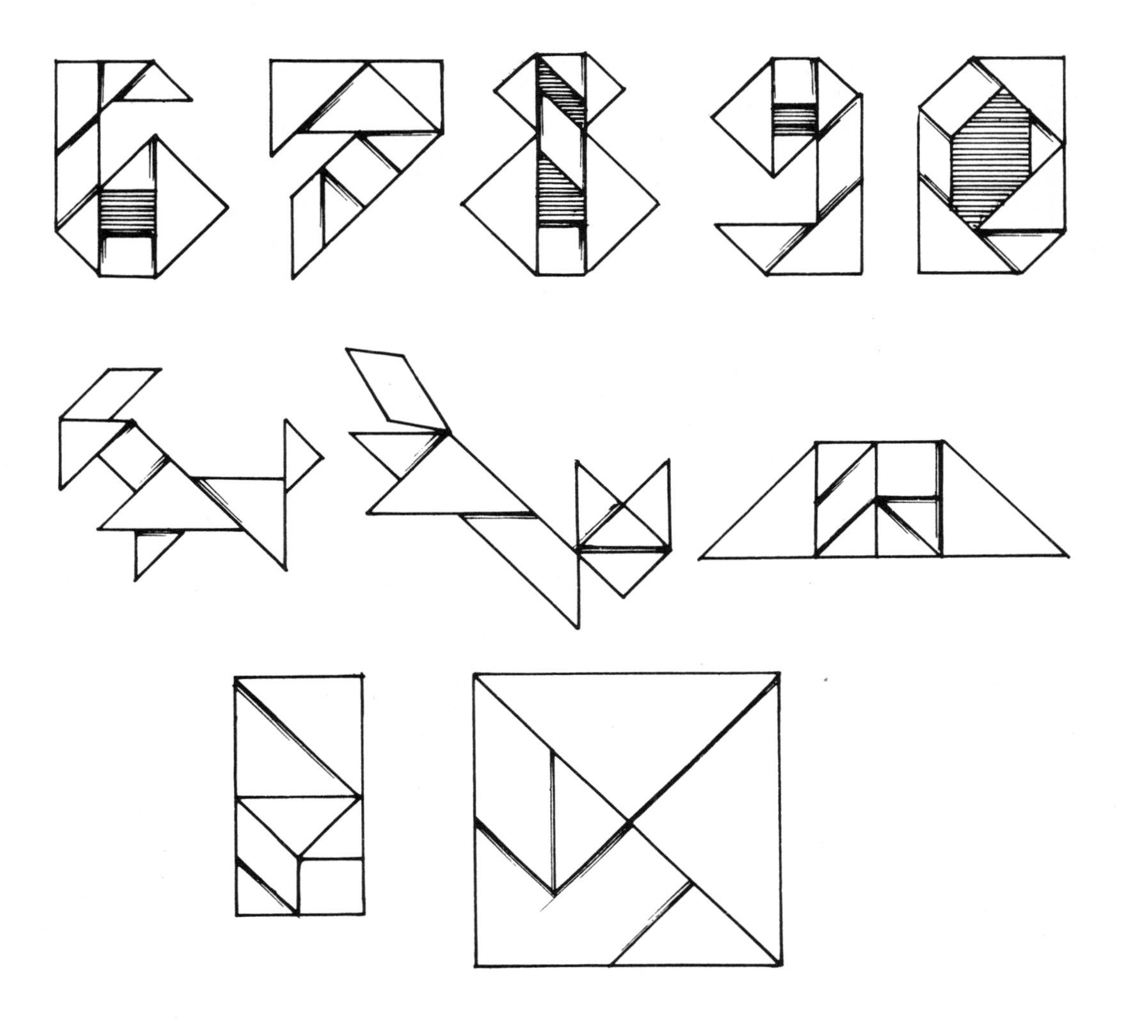

each set of alphabet blocks. Paint or varnish the wood pieces for a more finished product. See Figure 5.10.

OBJECTIVES:

1. To enhance the child's awareness of form perception.
2. To assist children in developing concepts relative to laterality, directionality, position in space, and spatial relationships.
3. To help children develop fine-motor coordination.
4. To develop concepts of balance.
5. To enhance development of visual and auditory sequential memory.

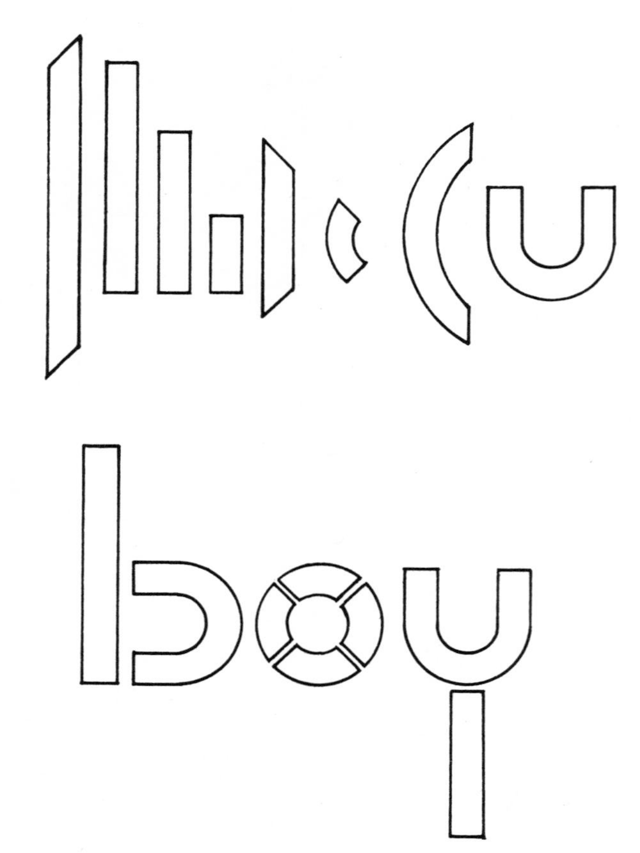

FIGURE 5.10. Alphabet blocks.

ACTIVITIES:

1. Stack the block pieces as high as possible, without letting them fall.
2. Arrange the block pieces into the letters of the alphabet.
3. Arrange the block pieces into numbers or pictures.
4. Use auditory directions to give children clues as to how to arrange the pieces.
5. Make a pattern, letter, or number with some blocks. Have the child look at your design. Then, remove the design from his sight and have him construct one from visual memory.
6. Have the child look at a design, close his eyes, and reproduce it through visual memory by using his tactile-kinesthetic sense.

* For further suggestions, refer to parquetry blocks in another section of this book (pp 48-49).

FORM PUZZLES

CONSTRUCTION:

This piece of equipment can be made from 5 x 8" index cards, construction paper, envelopes, and tape. Tape two index cards together (see Figure 5.11). Cut

FIGURE 5.11. Form puzzles.

two sets of puzzle pieces out of construction paper and glue one set to one of the index cards. On the second index card, draw an outline of the pieces. The second set should be placed in an envelope and glued to the back of one of the index cards. When a child wishes to use the puzzle, he can take the second set out of the envelope and try to match the puzzle pieces with the outline on the index card.

OBJECTIVES:

1. To assist children in developing form perception.
2. To assist children in developing concepts relative to position in space and spatial relationships.
3. To enhance visual and auditory sequential memory.
4. To develop auditory and manual encoding abilities in children.

ACTIVITIES:

1. Place the puzzle at the edge of a table, as seen in Figure 5.11. The completed puzzle should hang over the edge out of the child's sight. The outline of the puzzle should be on the table top. Have the child take the puzzle pieces out of the envelope and put the puzzle together by matching the pieces with the outlines. When the

child is finished with the puzzle, flip the model side up to see if it matches with the side that the child has just completed.

2. Just use the puzzle pieces without the index cards. Give auditory instructions to the child telling him how and in what order to make a design. For example, "place a yellow triangle to the right of a red square. Place a green diamond to the top of the red square."
3. After the child completes the puzzle, have him verbally describe to you the position or relationship of one puzzle piece to another.

6

Visual Perception

Visual perception is one of the receptive phases of language development. It is a complex phenomenon that should have the attention of a developmental vision specialist. As educators, however, there are many areas in which we can work in cooperation with an optometrist in order to meet the needs of each child. Prior to working in this area, it is essential to understand the difference between sight and vision. Sight is the ability to see. It is the eye's response to light shining into it. Vision is the result of the child's ability to interpret, understand, and define the information that comes to him through his eyes. Many children have good sight, and yet have critical and inhibiting vision problems that affect reading and other aspects of academic performance (2, 21).

Concepts relating to myopia, hyperopia, amblyopia, strabismus, astigmatism, exophoria, esophoria, and other pathological conditions require medical techniques and should be treated by a professional optometrist. Our concern, as educators, is with the child who has adequate ocular machinery, but who has not learned to use it efficiently. Educators may work with concepts relating to visual perception that directly concern visual tracking or pursuit, saccadic movements, convergence-reconvergence, visual figure-ground perception, and visual sequential memory (6, 21, 27).

It is essential for the child's eyes to work together in a coordinated fashion to perceive information efficiently. The child's ability to track visually a moving object is directly related to school tasks, such as catching a ball or watching a bird fly. As the child learns to read, the eyes must move in a saccadic fashion, from left to right, before refocusing on the next line. If the child learns to move the head while reading the stationary words on the page, an inefficient or splinter skill develops. The child's eyes must also be trained to converge in order to work at close vision while using a workbook or reading at a desk, and then quickly reconverge to distant vision by looking at the teacher or chalkboard. The child who cannot adjust efficiently will often loose his place and become confused. Training for visual figure-ground perception is essential because it relates to how well the child can distinguish the primary figure from the background, whether it be catching a ball or locating a word or picture on a page. Visual sequential memory is also an important aspect of perception because it enables a person to recall or remember what was seen.

MARBLE TRACK

CONSTRUCTION:

Marble tracks may be purchased commercially from one of the companies listed in the resource section at the end of the book. The price is usually quite reasonable, but if you have the tools, you can make one inexpensively. Purchase a piece of 1" x 1" wood, 10 feet long. Cut it in half the long way to make 2 pieces of ½" x 1" wood 10 feet long. Cut each of the 1-foot pieces into five equal pieces, 2 feet in length. By using small nails, attach two 2-foot pieces together to make a "V", as seen in Figure 6.1; do this with the remaining pieces. Make two side-support pieces out of a 10-foot piece of 1" x 2" wood. Cut the piece of 1" x 2" wood lengthwise to make two pieces of ½" x 2" wood, 10 feet in length. Cut each of the 10-feet pieces into five equal sections, 2 feet in length. By using small nails, assemble three 2-foot pieces together to make a "U", as seen in the Figure 6.1. With three more pieces, make a second side support. Use two pieces of 6 inch square 1/8" or 1/4" plywood to serve as bases. Nail the bases to the side supports. Attach the five tracks to the side supports with small nails or screws. Allow for a 10 to 15° slope on each of the tracks. Allow enough space between each of the tracks and side supports for a marble to fall to the next lowest level. See Figure 6.1.

OBJECTIVES:

1. To help children develop ocular pursuit abilities.
2. To develop eye-hand coordination.

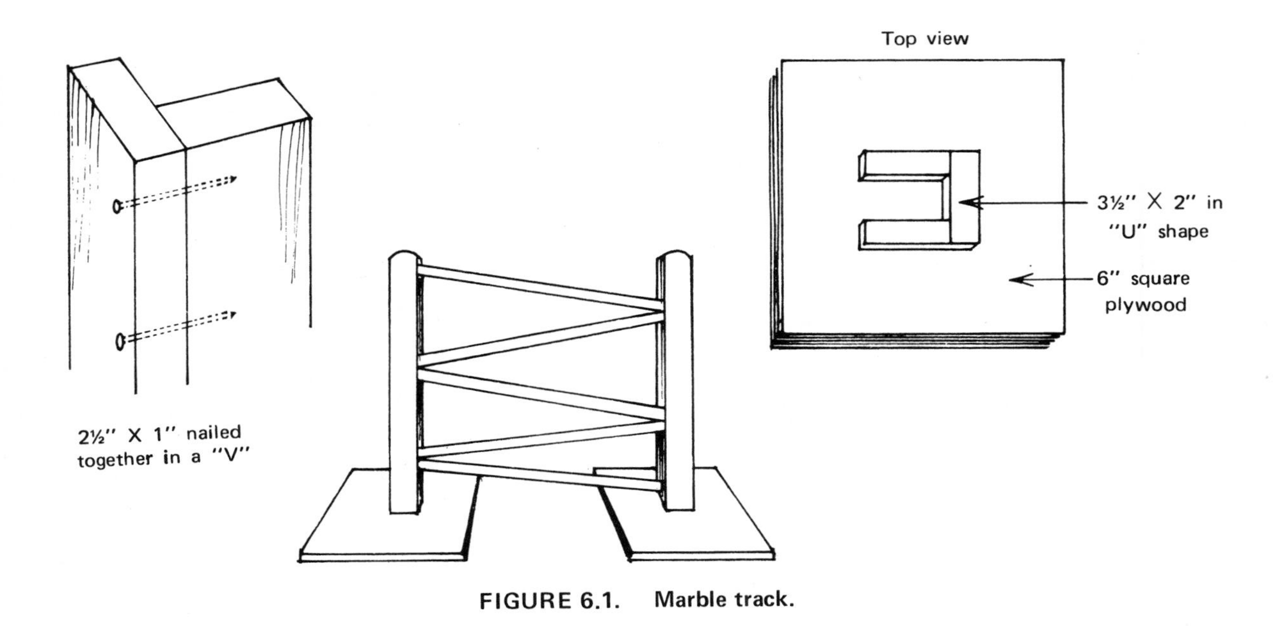

FIGURE 6.1. Marble track.

3. To orient children to the left to right and top to bottom progression used in reading.

ACTIVITIES:

1. Set the marble track on a table. Ask the child to sit at a chair in front of the table. Ask the child to watch the marble as it rolls down the track. He should keep his head still, moving only his eyes.
2. Ask the child to use one eye at a time. Occlude the second eye.
3. Have the child follow the marble with his eyes and his index finger from one hand as it rolls down the track. Alternate hands.
4. Have the child watch the marble as it rolls down the track and instruct him to stop it on command with his finger.
5. Have the child follow the marble down the track as the teacher tilts the track to simulate the movement of the eyes when reading; the marble rolls slowly to the right, then quickly to the left (saccadic fashion).

SWINGING BALL

CONSTRUCTION

A ball and a length of strong string or rope are needed to make this piece of equipment. The ball may be any size, from the rather large tether ball to the smaller plastic wiffle balls. The string or rope will vary in length, depending on whether the teacher will hold the swinging ball or whether the swinging ball will be permanently attached to the ceiling, as shown in Figure 6.2. In the latter case attach hook or eye mounts to the ceiling and side walls so that the swinging ball may be permanently suspended from the ceiling. When the ball is being used, it

FIGURE 6.2. Swinging ball.

may be lowered to the desired height. When the ball is not being used, it may be drawn up to the ceiling where it will not interfere with the other activities.

OBJECTIVES:

1. To help children learn patterns of visual pursuit.
2. To help children develop eye-hand coordination.
3. To assist children to visually respond to a moving target on a given command.

ACTIVITIES:

1. Have the child sit or stand 18 to 24 inches in front of the suspended ball. Pull the ball to the side and allow it to drop. The ball should swing from side to side using it's own momentum. The child visually follows the pathway of the ball. The child's head should not move.
2. Perform the latter activity by having the child follow the ball with one of his hands and with his eyes.
3. Move the child or the ball so that 4 to 5 feet separates them. Swing the ball in a forward-backward or to-fro direction. Ask the child to follow the path of the

FIGURE 6.3. Swinging target ball.

ball with his eyes, all the way in and back out again (convergence-reconvergence).

4. As the ball approaches the child, ask him to tap it alternately with both hands, first with the right, then with the left. Make sure that he taps the ball gently.
5. Swing the ball sideways again. Ask the child to remain in an alert position with his hands at shoulder level. As the ball swings from side to side, give the command "Now" at which time the child should immediately tap the ball. This task requires the child to follow the moving target at all times with his eyes, and to strike the ball with his left hand when it is on his left side and with his right hand when it is on his right side.
6. Give the child a rolling pin. At first swing the ball in a forward-backward direction. Have the child tap the ball with the rolling pin to keep it in motion.
7. Hang a suspended ball near a wall, as seen in the picture. Place a target on the wall. The target should have vertical zones painted in different colors, such as blue, red, green, red, and blue. Give the child the rolling pin and swing the ball from side to side. Ask him to trap the ball as it swings in the different color zones. See Figure 6.3.
8. As the child is sitting in a chair, move the ball in a circular pathway around his head at eye level. Ask the child to follow the ball across his field of vision and to move his eyes back so that he can see the ball again when it enters his field of vision. Move the ball in a clockwise and a counterclockwise direction so that the child can pursue the ball in a left to right progression and a right to left progression.
9. Ask the child to lie down on the floor above the suspended ball. Move the ball in a horizontal, vertical, diagonal, and rotary pathway and ask the child to follow it's path with his eyes. Only the

eyes should move. He should not move his head.

10. Have the child perform the latter activity, but ask him to follow the target with one of his hands, as well as his eyes.
11. Perform all of the previous activities with both eyes and then occlude one eye at a time to give the child a chance to practice with monocular vision.

METRONOME

CONSTRUCTION:

Because of it's versatility, a metronome is a valuable contribution to several aspects of a perceptual-motor program. However, it is difficult to make. It is therefore suggested that one be purchased from one of the companies listed in the resource section at the end of this book. A spring-monitored metronome, which is wound up, will cost between $10.00 and $20.00. An electric metronome will cost between $20.00 and $30.00.

OBJECTIVES:

1. To help children learn an awareness of rhythm.
2. To help children who have tendencies toward hyperactivity learn to pace their activities.
3. To help children learn visual pursuit.
4. To help children learn sequencing patterns.
5. To help children develop eye-hand and eye-foot coordination.

ACTIVITIES:

1. Clap your hands, snap your fingers, stamp your feet, or march to the rhythm of the metronome. (Most metronomes will range from 40 to 200 beats per minute, depending on the setting.) Begin with a slow tempo and, as the child gains confidence, increase the rhythm.
2. Follow the path of the meter stick, using only your eyes. Do not move your head.
3. Follow the path of the meter stick with your eyes and one index finger. Switch or alternate hands.
4. Perform physical fitness exercises to the rhythm of the metronome. For example, do jumping jacks to the rhythmical pace set by the metronome.
5. Create a simple locomotor or dance sequence and perform to the rhythm of the metronome. Include finger snaps, hand claps, and thigh taps with your sequence. Hops, walks, jumps, and runs will work well with the even rhythm of the metronome. Uneven rhythms, such as a skip, slide, or gallop will not work. Have the children make up a sequence individually or with a partner.
6. Perform sequences, such as the Look, Ready, Touch, and Back exercises to the rhythm of the metronome (pp. 97-98)

7. Perform the Chart Exercises involving the arrows, triangles, and fingers to the tempo of the metronome (pp. 47-48)

SACCADIC MOVEMENTS

CONSTRUCTION:

Saccadic eye movements are coordinated and allow the eyes to work together in moving or jumping from one word or phrase on a written page to another, as the progression proceeds from left to right and from the top to the bottom of the page. All that is needed for these activities is a metronome and two pencils or objects on which the child can concentrate his vision.

OBJECTIVES:

1. To enable the child to coordinate his eye movements.
2. To help a child fixate on a target and switch from one target to another in a convergence-reconvergence fashion.
3. To assist the child in developing rhythmical awareness.

FIGURE 6.4.

4. To help a child learn to follow visual and auditory cues.

ACTIVITIES:

1. Give the child two pencils, one for each hand. Name the pencils right-left, or one-two. Have him extend his arms at eye level and look at the erasers on the pencils. First have him look with both eyes at one pencil; then have him look at the second pencil. Continue alternating sides in a rhythmical fashion. See Figure 6.4.
2. Perform the latter activity; then, ask the child to vocally say right or left (one or two) as he looks with his eyes in that direction.
3. Perform the latter activity using a metronome to set the tempo.
4. Add motoric movement to the task by having the child step with the right foot as he looks and says "Right" and step with the left foot as he looks and says "Left."
5. Use opposition in the motoric movements by stepping with the left foot as the child looks and says "Right."
6. As the child performs the task at a refined level, begin asking him to concentrate on the whole visual field or background rather than only the pencil erasers. For example, ask him to name objects in the room as he moves his eyes from right to left.

EYE ROLLS

CONSTRUCTION:

The activities in this section require the use of a balance board, pie tin, and a marble. No construction is necessary.

OBJECTIVES:

1. To develop visual tracking abilities.
2. To develop balancing skills.
3. To help children develop spatial awareness concepts.

ACTIVITIES:

1. Have the child stand or sit up straight in a chair. Move your fingertip or pencil eraser through the child's visual field in a horizontal, vertical, diagonal, and rotary direction. The target should be 16 to 24 inches away from the child's eyes. The child's head should remain stationary while the eyes pursue the target.
2. While working in a rotary direction, have the child move his eyes in a clockwise direction. Then have him move in a counterclockwise direction.
3. Perform this activity while the child has his eyes closed, pretending that he is following a target in a rotary fashion.
4. Have the child stand on a balance board and follow a target as it is moved in a horizontal, vertical, diagonal, and rotary

FIGURE 6.5. Eye rolls.

direction. Emphasize the rotary movements. The child should stand still on the board in a balanced position, keeping his head still, and using only his eyes to follow the target.

5. Have the child stand on the floor in an erect position. Have him look at a stationary target held 16 to 24 inches away from his head. Have him rotate his head and eyes in a clockwise and counterclockwise direction while he maintains focus on the stationary target.
6. Perform the latter activity while the child is standing on a balance board.
7. Teach the child to use his index finger as the target to track in a horizontal, vertical, diagonal, and rotary direction. This will teach the child eye-hand coordination and space-time sequencing. Perform this activity first with the eyes open, and then, with the eyes closed. As the child has his eyes closed, be certain that his eyes are keeping pace with the moving target. You can see the pupils move under the eye lids.
8. Have the child hold a pie tin in a horizontal plane, as seen in the picture above. Place a marble in the pie tin and have the child rotate it in a direction corresponding to the direction in a vertical plane. The eyes should follow the target being moved

through this plane. Have the child vocalize the target's location at the same time by saying top, right, bottom, or left. See Figure 6.5.

9. Perform this activity while the child is standing on a balance board.

FLASHLIGHT

CONSTRUCTION:

Purchase a flashlight from the local hardware store. The light may range from the small penlight to the common household flashlight. Purchase an extra set of batteries and always keep them on hand. See Figure 6.6.

FIGURE 6.6. Penlight.

OBJECTIVES:

1. To help teach children visual pursuit.
2. To develop their eye-hand and eye-foot coordination.
3. To help children develop the concepts of fixation and convergence-reconvergence.

ACTIVITIES:

1. Hold a penlight 16 to 18 inches away from the child and move it in horizontal, vertical, diagonal, and rotary directions. Ask the child to follow the moving light with his eyes. There should be no head movement. Follow the target with both eyes, right eye only, and left eye only.
2. Perform this same activity and ask the

child to follow the light with his eyes and the index finger from one hand at the same time.

3. Perform these two activities while the child is lying down as well as sitting or standing.
4. Place two flashlights side by side, about 5 feet apart or one in front of the other, about 5 feet apart. Ask the child to fixate or converge his vision on one light and then, on command, switch his vision to the other light. The child should stand or sit from 5 to 10 feet away from the lights.
5. Turn off all the lights in the room and make the room as dark as possible by closing the door and drawing all of the window shades. Move the flashlight around the room or ask the child to shine the beam around the room. The child should concentrate on keeping his eye movement with the path of the light.
6. With a flashlight in each hand in a dark room, ask the child to shine the light on his feet. Begin by asking the child to walk and keep the light shining on his feet by moving from right to right and vice versa. Then, alternate right arm to left foot and vice versa.
7. Dim the light in a classroom full of various objects. Give the child a flashlight and ask him to find different objects in the room sequentially by shining the light on them.
8. Make a reading chart with large manuscript letters. Darken the room and give the child a flashlight. Ask him to find various words on the chart or to read from left to right and from the top of the chart to the bottom.

TACHISTOSCOPE AND SLIDES

CONSTRUCTION:

A tachistoscope is a camera lens that can be attached to a slide projector. It permits slides to be shown on the screen at different rates of speed from 1/125 to 1 second. It is often used to teach speed reading. Tachistoscopes can be purchased for $98.00 to $350.00 from one of the companies listed in the resource section of this book. This is expensive and may be prohibitive for many school systems. However, the same relative effect can be achieved by holding one's hand over the slide projector's lens and withdrawing it momentarily. Some of the speed and accuracy are lost, but at a rudimentary level of perceptual-motor development, these factors are not essential. The slides can be made by purchasing clear acetate 35mm film and empty 2 x 2" slide mounts at a camera store. Use a felt tip pen or grease pencil to draw on the acetate. Cut the acetate to the proper size and insert it into the slide mount. Make several different series of slides, as seen in Figure 6.7. Make approximately 10 slides for each series.

OBJECTIVES:

1. To develop visual memory.
2. To develop spatial awareness.
3. To develop form perception.
4. To develop eye-hand coordination.

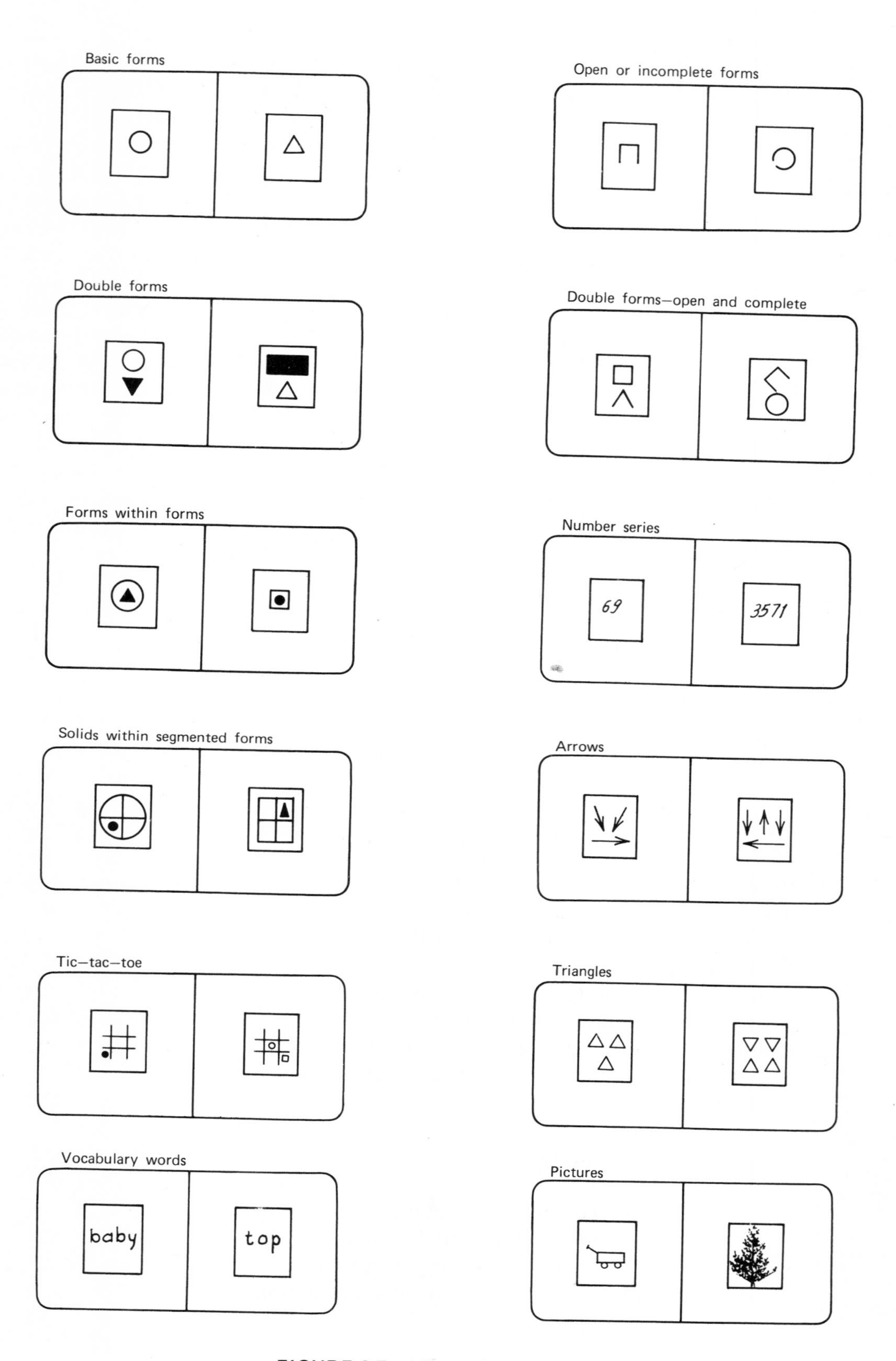

FIGURE 6.7. **Examples of slides.**

ACTIVITIES:

1. Tell the child you are going to flash a form on the screen. The command will be "Ready-Now." After he sees the image on the screen, ask him to close his eyes and visualize it in his mind's eye. Then ask him to open his eyes and tell you what he saw.
2. Repeat these same procedures, but this time, ask the child to trace what he saw in the air. (Kinesthetic)
3. Repeat the same procedures as above, but this time, ask the child to find and circle the same image on a worksheet. (Matching)
4. Performing the same procedures, ask the child to find and trace over the same image on his worksheet. (Tactile-Kinesthetic)
5. Repeat these same procedures, asking the child to make an image similar to the one he saw from memory on a blank piece of paper. (Visual Decoding-Visual Encoding)

** When using the tachistoscope and slides, a sequence of ten frames at each session falls within the attention span and fatigue level of the average child. Begin with a slow flash speed and increase the speed gradually. Start with the simple forms and then increase the complexity. Gradually, the child should become more discriminatory in terms of size, color, shape, similarities, differences, equals, unequals, laterality, directionality, spatial relationships, and so on, with reference to the five previous exercises. Each of these elements is important to symbolization and cognition. When a child has internalized each of these foundational concepts by making them personally meaningful, and can skillfully use them for further building, abstraction, and generalization, that child is ready for reading.

FLASH CARDS

CONSTRUCTION:

Make a series of simple patterns on 3" x 5" index cards. Use geometric designs. arrows, letters, numbers, tic-tac-toe, and domino designs, as shown in Figure 6.8. Use combinations of singles, twos, and threes of each item. Also use combinations of the same size, but different shapes; use same shapes, but different sizes; and use same shapes, but different orientation, and so on. See Figure 6.9.

OBJECTIVES:

1. To assist the child in utilizing his visual memory to recall accurately the size, orientation, and shape relationships in a quick exposure.
2. To enable the child to manipulate and reproduce patterns in an orientation that differs from the one presented.
3. To enable a child to decode and encode messages quickly.

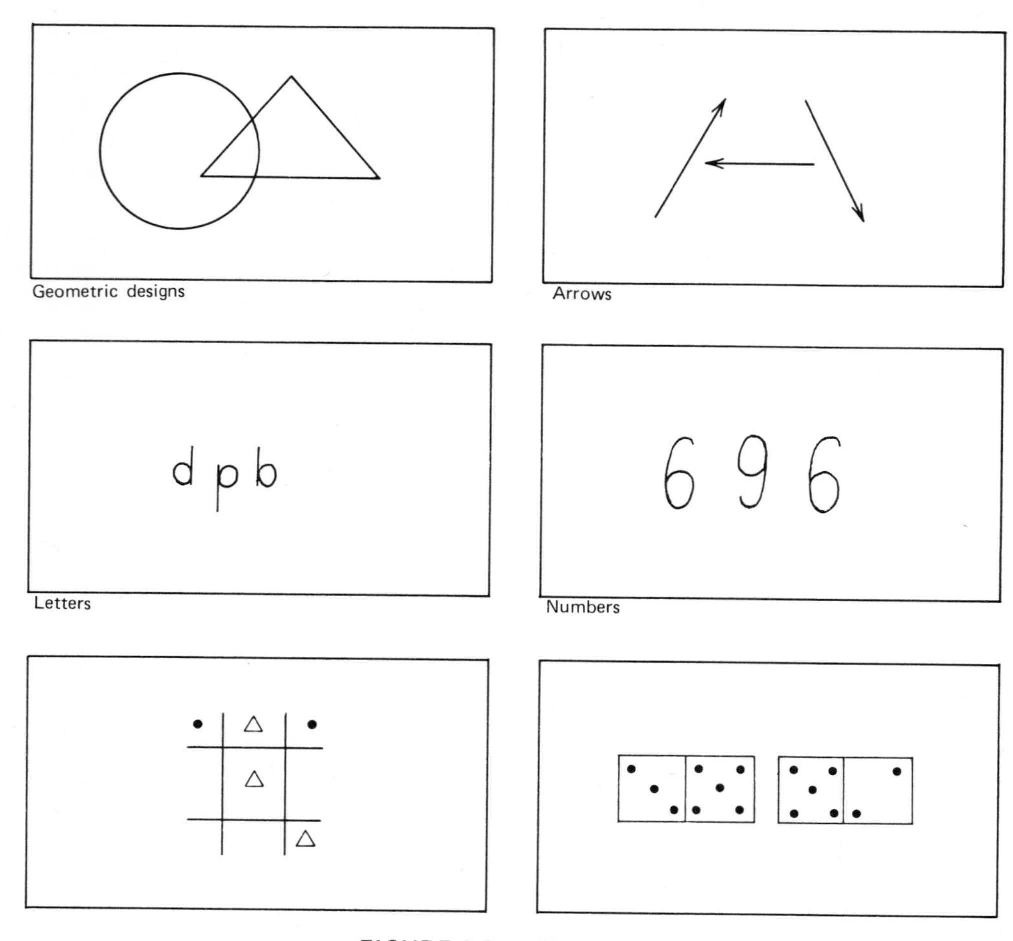

FIGURE 6.8. **Flash cards.**

ACTIVITIES:

1. The teacher should hold the card away from the child's view or occlude the card. On the command, "Ready-Now," the teacher should momentarily expose the card to the child. The child should perform the following activities in sequence:
   a. Name or identify what he saw.
   b. Trace what he saw in the air.
   c. Identify a similar pattern on a worksheet.
   d. Draw the same pattern on a chalkboard or piece of paper.

* By rotating each card, four different exposures are possible. Show the card to the child, after each exposure, for immediate feedback.

2. Show one card to the child. Then show the child a similar card with one or two aspects changed and ask him to tell you what is missing or what is changed.
3. Expose various patterns to the child and ask that they be reproduced as though seen with a ¼ or ½ clockwise or counterclockwise turn from the original position. See Figure 6.9. Start with a simple pattern, such as an arrow, and move toward more complex activities.
4. Play tic-tac-toe using the flash method. Have two children play or have the instructor play with the child. The first person

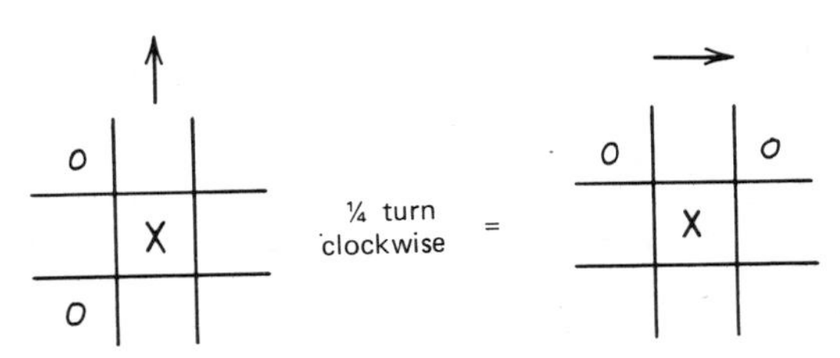

FIGURE 6.9. Rotated orientation.

starts the game by making a mark, and then flashes his card to his opponent. The opponent, in turn, adds his mark and returns the flash until the game is completed.

5. Set up a tic-tac-toe situation on a flash card. Instruct the child to place an *X* or an *O* in the most appropriate square, either to win or to block the opponent from scoring.
6. Perform flash-matching experiences with parquetry blocks. Make a pattern out of three or more blocks. Occlude the pattern from the view of the child. On the command, "Ready-Now," momentarily expose the pattern to the child. Have the child draw the same pattern on a piece of paper or have him make it with the blocks. Rearrange the pattern by shifting one block, and then repeat the experience. Ask the child to rearrange his pattern with a minimal amount of block shifting to match what he saw. Increase the difficulty as the child progresses.

LOOK, READY, TOUCH, AND BACK

CONSTRUCTION:

The equipment for this exercise can be readily obtained from common household items. Two toys, two playing cards, or two objects can be used. If a metronome is available, it could be used for rhythmical sequencing of activities. See Figure 6.10.

OBJECTIVES:

1. To develop ocular fixations, visual pursuit, convergence, and reconvergence.
2. To help the child learn to plan motorically (manual encoding) the activities he performs.
3. To develop visual and auditory sequencing patterns.

ACTIVITIES:

1. Two visual targets should be placed three feet apart on a wall or chalkboard at the child's eye level. The child should stand approximately three feet away from the targets.
   a. The child should look at the target on the right and call "Look" on the first beat of the metronome.

FIGURE 6.10. Look, Ready, Touch, Back.

b. On the second beat of the metronome, he should call "Ready," as he raises his right hand to touch the target.
c. On the third beat, he should call "Touch," as he reaches out to touch the target on the right.
d. On the fourth beat, he should call "Back," as he returns his right hand to his side.
e. Perform the same activities with the left hand.
f. Perform these activities and step forward with the right foot when the right arm moves and step forward with the left foot when the left arm moves.
g. Move the left foot with the right arm and the right foot with the left arm.
h. Repeat the activities without the auditory support of saying, "Look, Ready, Touch, Back." Do it silently in time with the metronome.
i. Repeat the activities, and instead of calling out the words, have the child tell a story, such as the plot of a book or a TV program, something that might have happened during the day, or an account of a trip or visit. Keep in time with the metronome.

EMBEDDED FIGURE PUZZLES

CONSTRUCTION:

Pictures, such as Figure 6.11, can often be found in children's magazines, on restaurant placemats, or in children's books. You can also use any picture that contains a scene or perhaps even draw your own pictures. After you have found the pictures, have them dry mounted or cover them with a sheet of clear plastic so that the children can draw on them with a grease pencil, erase, and use them again.

FIGURE 6.11. Hidden pictures. Can you find these hidden objects in the large picture above?

fork	feather	mouse	toothbrush
heart	thimble	kettle	safety pin
fish	letter Z	shovel	clothespin
tepee	jacknife	apple	fish hook

OBJECTIVES:

1. To help children develop better visual-figure-ground perception.
2. To help children develop an organized pattern of visual search.
3. To develop form perception and form constancy.

ACTIVITIES:

1. Try to find each of the objects in the picture by yourself or with a partner.
2. Try to draw each of the objects that you see in the picture on a separate sheet of paper.
3. Can you draw your own pictures with embedded figures? Make a picture for a partner to work with. Take turns.

BUTTONS

CONSTRUCTION:

Collect as many different types of buttons as possible and store them in a convenient container. See Figure 6.12.

FIGURE 6.12. Sorting buttons.

OBJECTIVES:

1. To help children develop visual figure-ground perception.
2. To help children learn to classify objects

according to size, shape, color, texture, and so on.
3. To develop eye-hand coordination.

ACTIVITIES:

1. Scatter the buttons on a plain surface. Have the children sort the buttons according to color, size, texture, number of holes, shape, and so on.
2. Perform the same activity, but use background surfaces that offer more distractions, depending upon the child's ability.
3. Arrange the buttons into various geometric shapes, letters, numbers, or simple pictures.
4. While sitting, kneeling, or standing, drop buttons into a container with a small opening. For example, drop the buttons into the mouth of a plastic bottle.
5. Flip the buttons with your fingers, as you would a coin.
6. Play a game similar to Tiddly Winks with the buttons.
7. Shoot the buttons with your fingers, as if you were playing marbles or backgammon.

PAPER AND PENCIL ACTIVITIES

CONSTRUCTION:

There need not be any construction, but if one wishes to concentrate on specific letters, it may be advantageous to select words of specific interest, cut them out, and paste them all on one page. Under these conditions, one may also present various backgrounds (black on white, white on black, and so on) as an added incentive. Otherwise, any given page out of a newspaper, magazine, or workbook will suffice. See Figure 6.13.

OBJECTIVES:

1. To develop concepts of spatial awareness, including laterality, directionality, midline, position in space, and spatial relationships.
2. To develop visual figure-ground perception.
3. To enhance the development of a systematic process of scanning from left to right and from top to bottom.
4. To develop fine-motor coordination.

ACTIVITIES:

1. Find all the words on the page that begin with "p". Draw a circle around all of them.
2. Find all the words on the page that begin with "b". Draw a straight line under each of them.
3. Find all the "o's" on the page and fill in the hole with your pencil.

FIGURE 6.13. **Find the words that begin with *b* and *p*.**

4. Draw a connecting line from one "a" on the page to another, until you have connected all of them. Repeat this procedure with other letters.

MAZE BOX

CONSTRUCTION:

Get a cardboard box that is approximately 18 x 24" in dimension, from a grocery or hardware store. Empty soap cases, vegetable cases, canned fruit cases, and soft drink cases make good containers. Cut the box so that it is 4 to 6 inches deep (18" x 24" x 4"). Use manila folders to construct walls inside the box (see Figure 6.14).

OBJECTIVES:

1. To help the child develop organized patterns of visual search.
2. To develop eye-hand coordination.

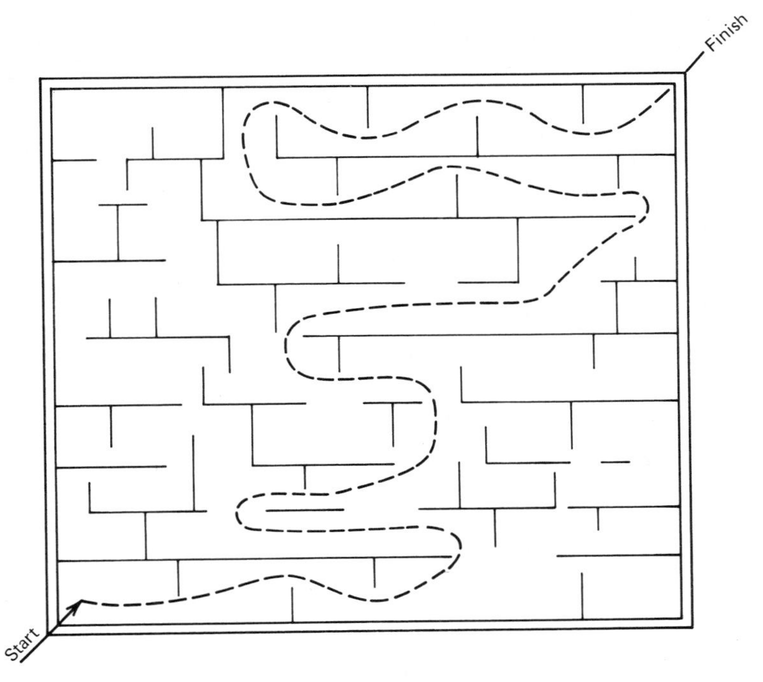

FIGURE 6.14. Maze box

ACTIVITIES:

1. Have the child place his index finger at the beginning of the maze. Instruct him to trace his finger on command, from the beginning to the end of the maze.
2. Place a marble in the box at the start of the maze. Give the child a pencil. By holding the eraser end down, have him use the pencil to manipulate the marble to the end of the maze.
3. Turn the box to the right, left, up, and down, so that there are four different starting positions for the maze. Turn the box on edge for additional positions.

WHAT IS MISSING?

CONSTRUCTION:

Develop a booklet of pictures or drawings. Each one must have a missing element. Examples of pictures with elements missing can be seen in Figure 6.15. Make up your own additional ones. Draw the pictures on 4 x 6" index cards or 8½ x 11" typing paper. Laminate each of the pictures with dry mount paper or clear plastic acetate so that the children can draw on the pictures without damaging them. The pictures can then be used repeatedly after erasing the children's marks. Keep the pictures in a notebook.

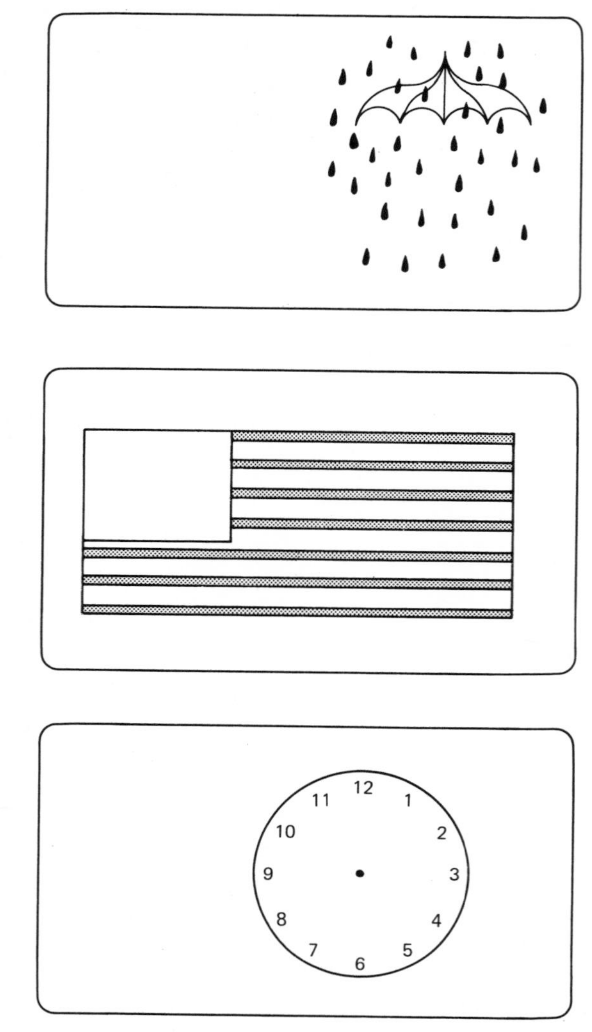

FIGURE 6.15. What's missing?

OBJECTIVES:

1. To help children develop organized patterns of visual search.
2. To help children learn form perception.
3. To help children learn to visually decode messages.

ACTIVITIES:

1. Page through the booklet and identify each of the missing parts in the pictures.
2. Use a crayon or a felt tip marker and draw each of the missing parts.

LADDERS

CONSTRUCTION:

Wooden or aluminum ladders, 8 feet long, should be painted in striped and solid fashions, as shown in the following pictures. Black and white are the suggested colors; however, other colors may be used. Sheets of 4 x 8' plywood should be painted in striped and solid designs to serve as the background. Other patterns may also be chosen. See Figures 6.16, 6.17, 6.18, and 6.19.

OBJECTIVES:

1. To develop visual figure-ground perception.
2. To develop better balancing abilities.
3. To develop better locomotor patterns in terms of eye-foot coordination.

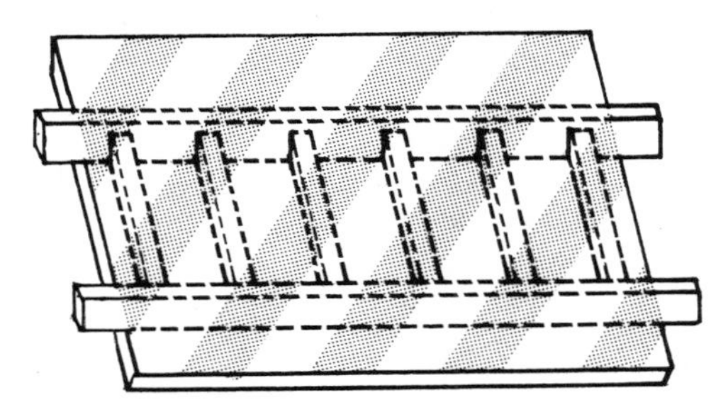

FIGURE 6.16. **Striped ladder—striped floor pattern.**

Ideas for the figure-ground ladders can be attributed to Dr. David Gallahue, Associate Professor, Indiana University, Bloomington, Indiana

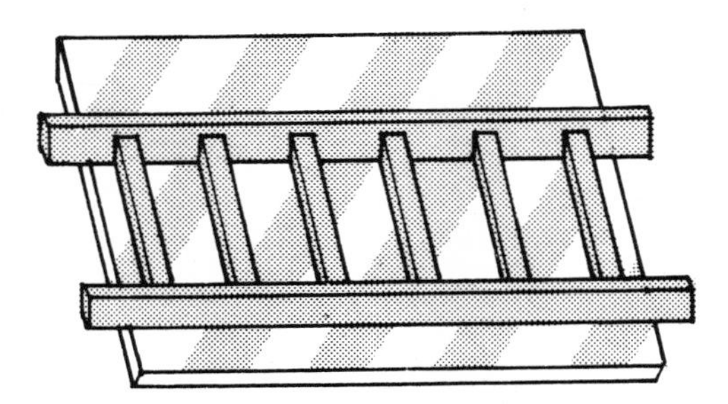

FIGURE 6.17. **Solid ladder—striped floor pattern.**

Ideas for the figure-ground ladders can be attributed to Dr. David Gallahue, Associate Professor, Indiana University, Bloomington, Indiana

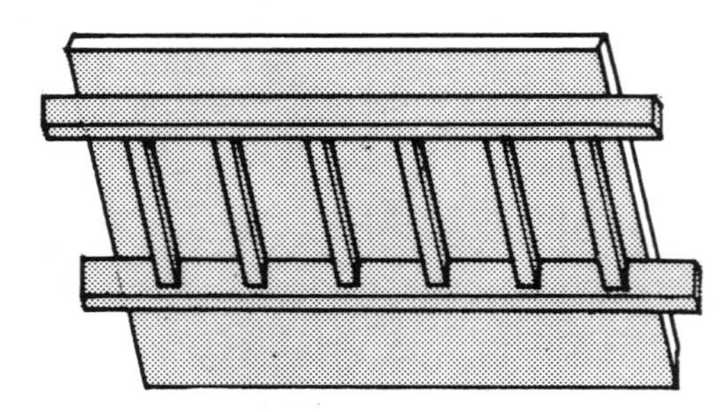

FIGURE 6.18. Solid cobred ladder—solid cover floor pattern.

Ideas for the figure-ground ladders can be attributed to Dr. David Gallahue, Associate Professor, Indiana University, Bloomington, Indiana

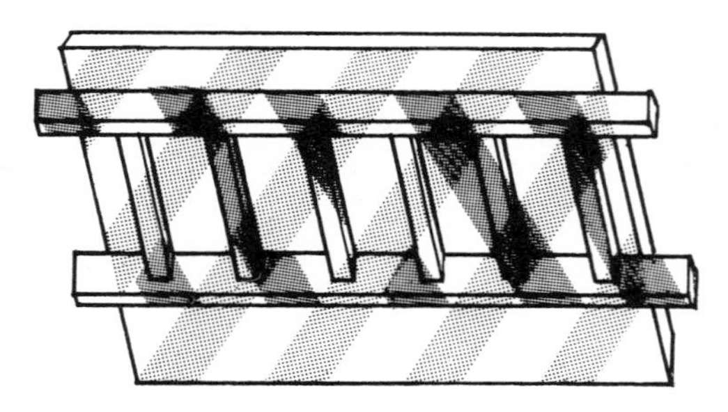

FIGURE 6.19. Stripped floor pattern—striped ladder in opposite directions.

Ideas for the figure-ground ladders can be attributed to Dr. David Gallahue, Associate Professor, Indiana University, Bloomington, Indiana

ACTIVITIES:

1. Walk from one end of the ladder to the other by placing one foot in between each of the rungs.
2. Hop with one foot from one end of the ladder to the other by placing the hopping foot in between each of the rungs.
3. Jump with two feet from one end of the ladder to the other by placing both feet in between each of the rungs.
4. Walk from one end of the ladder to the other by placing one foot on each of the sides of the ladder.
5. Walk sideways across the ladder by placing your feet on one of the sides of the ladder and sliding your feet.
6. Walk, hop, and jump from one end of the ladder to the other by moving in a backwards direction.
7. Alternately cross the midline of the body by crossing the feet from one side of the ladder to the other as you move from one end to the other.

*It has been found that children perform better where there is a distinct difference between the figure and the background. Therefore, begin the latter sequences with a figure-background relationship using contrast, and then, move to the more difficult patterns.

BALLS

CONSTRUCTION:

Scraps of yarn, ladies' nylon hose, paper, and socks can be used to make various types of balls used in manipulative activities. Because these balls are soft, they help children overcome their fear of the ball while simultaneously developing ball-handling skills. Yarn balls should be made with rug yarn. It takes from one to two skeins of yarn to make one ball, depending on the size of the ball desired. To begin constructing a ball, cut two cardboard circles, 4 to 6 inches in diameter, with a 2 inch hole in the center. Put the cardboard circles together. Wrap the yarn around both circles, through the hole and around, until the 2 inch hole is completely filled. Using a razor blade, cut the yarn between the cardboard circles all the way around. Place a strong piece of string between the circles and tie it. Cut and remove both pieces of cardboard. Paper balls are made by crumbling up pieces of newspaper or pages of magazines and then using masking tape to retain the round shape. Nylon balls are made by stuffing ladies' nylon hose into old socks until the desired size is achieved. Make sure that the nylons are stuffed in snugly so that the ball becomes resilient. Tuck the loose ends of the sock inside the ball and sew the opening shut. See Figure 6.20.

Various types of commercially produced balls, such as playground balls and wiffle balls, are necessary for bouncing and dribbling activities.

OBJECTIVES:

1. To develop familiarity with balls in a variety of situations.
2. To develop visual figure-ground perception.
3. To develop eye-hand and eye-foot coordination.
4. To develop fine-motor coordination.
5. To develop creativity in the use of balls.
6. To develop sport and game-playing skills.

ACTIVITIES:

1. Movement exploration.
   a. Find a spot by yourself and see what you can do with the ball while staying in your own space.
   b. Try using different body parts to move the ball and stay in your own space.
   c. Now, move the ball around the room anyway you would like, keeping the level of your body and the level of the ball the same and change these levels frequently.

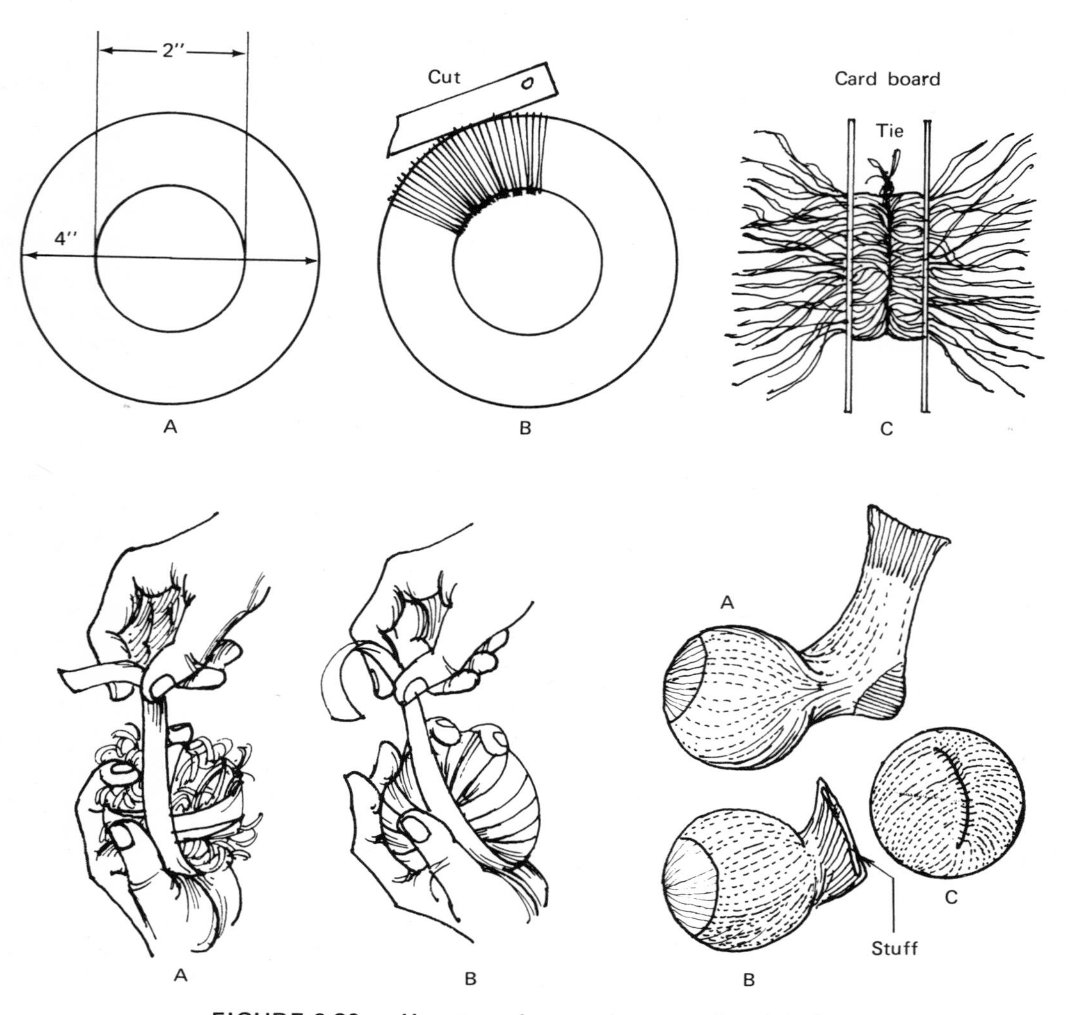

FIGURE 6.20. **How to make yarn, paper, and sock balls.**

d. Use many different body parts this time to move the ball around the room, keeping the ball close to you.
e. How many ways can you make the ball go around you?
f. What can you do with the ball, using one hand? Now try the other hand.
g. What can your feet do with the ball?

2. Rolling and fielding.
   a. See how many ways you can roll the ball around you, changing your base of support often.
   b. How many different directions can you use to roll the ball to yourself?
   c. Roll the ball in a straight line, run after it, and pick it up.
   d. Roll the ball, run after it until you get ahead of the ball, turn around, and pick it up. Try to keep your feet in motion.
   e. Roll the ball, leap over it, turn around, and pick it up by letting it roll up your arms.
   f. Roll the ball against the wall as hard

as you want. After it hits, run to meet it, and pick it up.

g. Roll the ball to a partner, but challenge him by making him move to field it.

h. In stride position (facing same or opposite directions), roll and field the ball between partners.

3. Throwing and catching.

a. Throw the ball to yourself, showing how many different ways you can throw and catch it.

b. How high can you throw the ball and still catch it? How did you make it go higher? Where should you be standing in order to catch the ball easily when it comes down?

c. How many different ways can you throw the ball without using your hands?

d. Can you throw the ball up and catch it before it bounces? How many different ways can you move and still catch it?

e. How many places can you put your hand and catch the ball?

f. Starting in a sitting position, throw the ball up and catch it while you are standing. Do the reverse.

g. How do you catch the ball most easily, with your arms bent or straight? Should you bring the ball in toward your body?

h. Can you throw the ball up and move somewhere to catch the ball before it bounces?

i. Can you throw the ball up in the air, turn around, and catch it before it bounces on the floor?

j. Can you throw the ball up in the air and clap two or more times before you catch it?

k. Standing in your own space, see if you can throw the ball behind you and catch it with the same hand.

l. Throw the ball up and catch it without making any noise. As soon as you make contact, give with the ball.

m. Hold the ball out to the side with your arm stretched. Bring the ball over your head and bring the other hand up to meet it to bring the ball down.

n. How many different ways can you get the ball from where you are, to the wall?

o. Try to throw the ball with your feet together. Now try it with your feet apart. Try it again, but this time with a step. Which way did you make your best throw?

p. I see we don't all step out on the same foot. Which foot do you use? Try it with both feet several times. Which way felt best?

q. What direction is your body facing when you start to throw? Does it make any difference in how far or accurately you can throw the ball?

r. Some of us are making our arms follow the ball after it is released and some are not. Try both ways and see which way is best.

s. Throw and catch a ball with a partner. Can you catch it when it is thrown to you? To your right or left? Can you move forward or backward to catch the ball?

t. Play catch with a partner, with each person having a ball. Now you have to concentrate on throwing your ball and keeping your eyes on your partner's ball so that you can catch it at the same time.

u. Try to juggle two or three balls. Perform the same activity with rings.

4. Bouncing and dribbling—playground balls are suggested for bouncing.

a. See if you can bounce the ball with

both hands with a lot of force.

b. See if you can control the force so that the ball bounces about waist high each time.
c. How many times can you bounce the ball in succession? Try bouncing the ball first with both hands. Now try to bounce the ball with just one hand.
d. Can you bounce the ball best with your right hand or left hand? Can you switch hands without losing control of the ball?
e. How should you use your wrist when you bounce the ball? Is it stiff or does it bounce with the ball? Which way can you bounce the ball the fastest?
f. Can you bounce the ball once and turn around before catching it again? Who can turn around twice?
g. How low can you bounce the ball? How high can you bounce the ball? Can you bounce it low, then high?
h. At how many levels can you be while bouncing the ball? Who can be on his stomach while bouncing the ball? Who can sit or be on his back while bouncing the ball? Can you get up while still bouncing it?
i. See how many different parts of your body can support you while you are bouncing the ball.
j. In how many directions can you move about the room while you dribble the ball?
k. Who can run, hop, skip, slide, or gallop while dribbling the ball?
l. When you move in different directions and dribble, where do you look or focus your eyes?
m. How many different parts of your body can you use to bounce the ball?
n. Can you bounce up and down as the ball bounces up and down?
o. Try to write your initials on the floor with the ball by dribbling in the shape of letters.
p. Get a partner and see how many different ways you can bounce the ball to your partner.
q. Can you bounce the ball to a partner while you move about the room?
r. See how you can make your body go over the ball as you give it a bounce with a great deal of force.
s. See if you can get under the ball in different ways as it is suspended in the air. Try different ways of getting your whole body or body parts under the ball as it is in the air.

5. Volleying.
   a. See if you can keep the ball in the air by using different parts of your body to keep it up. You may let it bounce on the floor between contacts, but try not to catch it.
   b. Choose two body parts and see if you can keep the ball in the air by alternating between them.
   c. Try using parts of your arms or hands to keep the ball in the air.
   d. Now see if you can keep the ball in the air using just your hands and fingers.
   e. Find a space on the wall, and see if you can keep the ball rebounding off the wall without catching it.
   f. See if you can volley a yarn ball in the air with forehand and backhand strokes while using an improvised nylon hose racket.
   g. Volley a yarn ball back and forth with a partner while using an improvised nylon hose racket.
6. Foot dribbling and kicking.
   a. See if you can move the ball with your feet without letting it get away from you.
   b. Try using different parts of your feet to move the ball. Use the sole,

toes, inside, and outside of your feet. Which part of your foot controlled the ball best?

c. Choose any part of your foot and see if you can change directions as you move the ball around the room.

d. Kick the ball with a medium amount of force and follow it. Try to stop it with your foot, without using your hands.

e. Kick your ball against the wall with any part of your foot and try to make it come directly back to you. When it returns, stop it with your foot or leg. Which part of your foot is most effective for kicking and stopping the ball?

f. Keeping the ball on the ground, how far can you kick it?

g. Some children are letting their foot follow the ball after they kick it, while some are stopping their foot when it meets the ball. Try both ways and see which makes the ball go farther. Which feels the best?

h. What did the foot that did not kick the ball do? What did your arms do? How does this help you kick?

i. Everyone, except four people get a ball. Moving your ball around the room, keeping it close to you, try to keep the students without a ball from getting yours.

j. With a partner (or in groups of four to five), take one or two balls and make up a game that involves sending the ball back and forth with various body parts.

7. Self-testing activities related to ball handling.
   a. Throw at targets on the wall using overhand and underhand throws for accuracy.
   b. Draw a 3-foot square or circle on the floor. While standing in it, throw the ball up high on the wall and catch it on the fly while you remain in the square or circle.
   c. Make two 3-foot squares on the floor, about 10 feet apart, in the same direction from the wall. Throw the ball on the wall from the first square so that you can catch the rebound while in the second square.
   d. Dribble and pass the ball through hula-hoops.
   e. Roll the ball, attempting to knock down Indian Clubs.
   f. Hit the ball against the wall, using one or two hands, overhand or underhand. Try side arm with one hand.
   g. Throw or kick the ball for distance.
   h. How many times can you hit the ball against the wall with a paddle without missing the ball?
   i. Dribble around pins in a zigzag manner. Do this with the feet also.
   j. Dribble under bars or ropes placed at various levels.
   k. Walk on a balance beam and bounce and catch the ball with each step.
8. Other ball-handling activities.
   a. Danish ball rhythms.
   b. Various games, relays, and sports.

GOLF TEE BOARDS

CONSTRUCTION:

Equipment needed for construction is a 16 inch square piece of ¼ inch plywood, a 16 inch square piece of pegboard, enough golf tees to fill every other hole, and four wood screws. Place the golf tees into the pegboard, as shown in

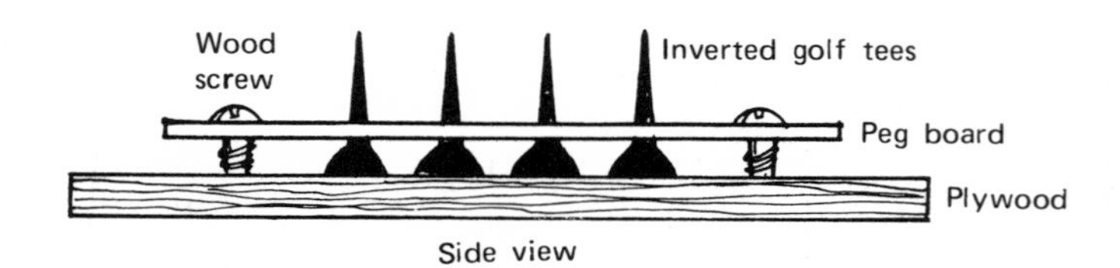

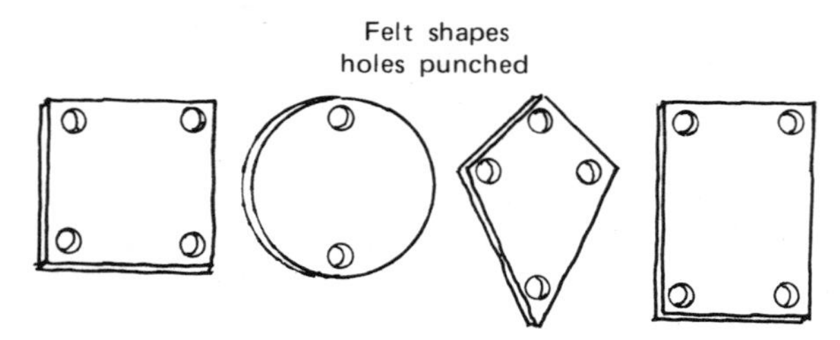

FIGURE 6.21. Golf tee boards.

Figure 6.21 and attach the plywood base. Make geometric shapes out of feltpaper, construction paper, or a similar material. Place holes in the shapes so that they can be placed on the golf tees.

OBJECTIVES:

1. To enhance visual figure-ground perception.
2. To develop concepts of spatial awareness including laterality, directionality, position in space, and spatial relationships.
3. To strengthen a child's concepts of form perception and form constancy.

ACTIVITIES:

1. Make a design, with different colors and shapes, on one board. Have the child reproduce the pattern on another board with the same spatial arrangement involving colors and shapes.
2. Make a design on a task card and have the child reproduce the same design on the board. See Figure 6.22.
3. Give auditory instructions directing a child to place various shapes and colors on the board with reference to spatial awareness.
   a. Place a red square, in the upper left hand corner.
   b. Place a blue diamond in the middle of the board, so that the long sides are pointing up and down.
   c. Place a yellow circle over the diamond so that the bottom half of the diamond is covered.
4. Write task cards instructing a child to place various shapes and colors on the board with reference to spatial awareness.
   a. Place a blue circle in the middle of the board.

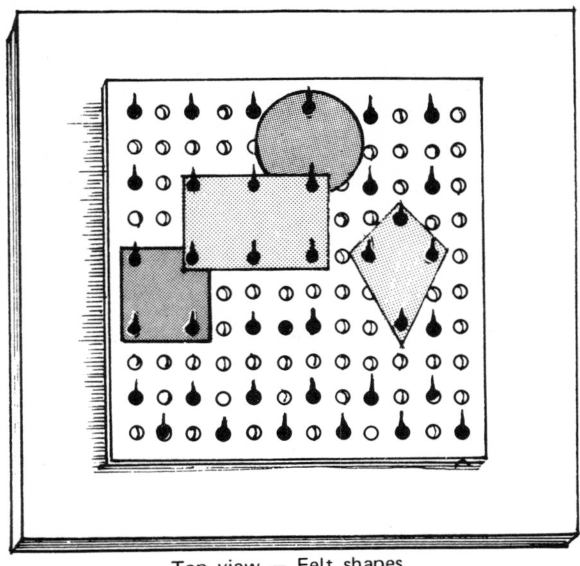

FIGURE 6.22. Make a pattern.

b. Place a red diamond to the right of the blue circle, so that the long sides are pointing sideways.

c. Place a green square in the middle of the two objects so that half of the blue circle and half of the red diamond are overlapped.

STRINGING BEADS

CONSTRUCTION:

An inexpensive commercial set of beads can be purchased containing beads of varying sizes, shapes, and colors. A shoe string is used to string the beads. Children also gain a great deal of enjoyment out of stringing popcorn or a collection of buttons. But, because the holes are nonexistent or very small, a needle and thread have to be used when stringing popcorn or buttons. In this instance, emphasize the importance of safety to the children. See Figure 6.23.

OBJECTIVES:

1. To assist children in developing visual memory and sequencing abilities.
2. To help children learn how to discriminate between objects varying in size, shape, and color.

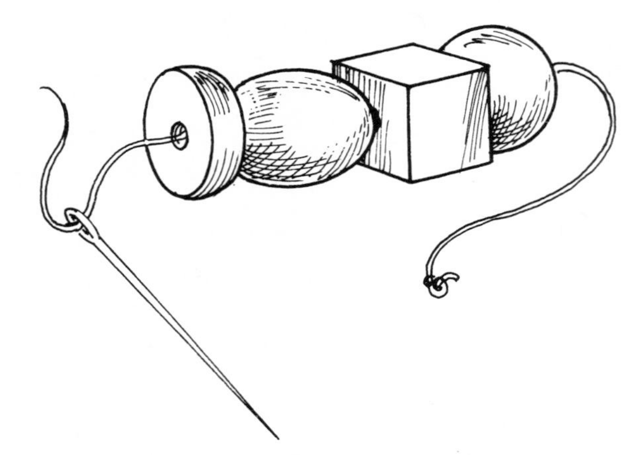

FIGURE 6.23. **Reproduce the bead sequence.**

3. To develop fine-motor coordination.
4. To develop tactile-kinesthetic awareness.

ACTIVITIES:

1. String a sequence of beads emphasizing one aspect, such as color (red, blue, green, yellow, red and so on). While the child is looking at the sequence of beads, give him an empty string and ask him to reproduce the same sequence. Perform the same activity using size or shape as the aspect to be emphasized. Use buttons or popcorn in the same fashion.
2. Add more than one dimension to the sequencing of beads. For example, work on color and shape at the same time.
3. In the number 1 exercise, ask the child which bead comes next in sequence (red, blue, green, yellow, red-*blue* . . .).
4. Show the child a sequence of beads, buttons, or popcorn containing three or more objects. After a momentary glance at the sequence, occlude it from his vision and ask him to reproduce it from memory.
5. Have the child close his eyes. Verbally tell the child a pattern. Then have him find the forms and place them on his string.
6. While a child has his eyes closed, allow him to feel a pattern on your string. Have him create the same pattern on his string.
7. String as many beads in as short a time period as possible. Keep time and have races to see if the child can improve his own score.

SEQUENCING BLOCKS

CONSTRUCTION:

A piece of scrap lumber from the local lumber supply company will be needed for this piece of equipment. A piece of 4 x 4", 1 x 6", or a piece of similar dimension will work. Cut the piece of wood into 4 to 8 slices, as shown in Figure 6.24. Each slice should be ½ to 1 inch thick. Sand each of the pieces to make

FIGURE 6.24. Sequencing blocks.

them smooth. Find four pictures that you wish to place on the sequence block. Pictures of animals, athletes, trucks, and toys, make good selections because they attract children. Cut the pictures into pieces the thickness of each slice of wood. Glue the pieces of picture to the pieces of wood. Smooth out the pictures well so that there are no air bubbles between the picture and the wood. Varnish over the picture on the wood and sand it lightly. Add several more coats of varnish sanding lightly after each coat. Make sure that each picture is glued to the wood in different sequence so that the child cannot put the sequence puzzle together once and have each side solved. A number code can be painted on the sides of the wood to assist the children in determining which puzzle pieces are in order.

OBJECTIVES:

1. To develop visual sequential memory.
2. To develop eye-hand coordination.
3. To assist children in learning how to balance a stack of blocks.

ACTIVITIES:

1. Scramble the puzzle pieces or sequence blocks on the table or floor in front of the child. Ask the child to put the blocks in sequence to make a picture.
2. Challenge the child to stack the blocks so they are in a balanced position.

SEQUENCING PUZZLES

CONSTRUCTION:

Use a large piece of corrugated cardboard or stiff poster board. With a knife, cut out a circle with a diameter from 18 to 24 inches. Cut out an inner circle with a diameter of 12 to 18 inches so that the product looks like a doughnut, as seen in Figure 6.25. Draw pictures using a central theme, such as toys or animals, around the remaining circular track. Draw from four to ten pictures, depending upon the complexity you desire. Cut the circular track into pieces, as shown in the picture, so that part of each picture is on two puzzle pieces. Use a coding system by numbers, shapes, or colors to label alternately each puzzle piece. Every other piece should be an odd or even number, a circle or triangle, or a red or green piece.

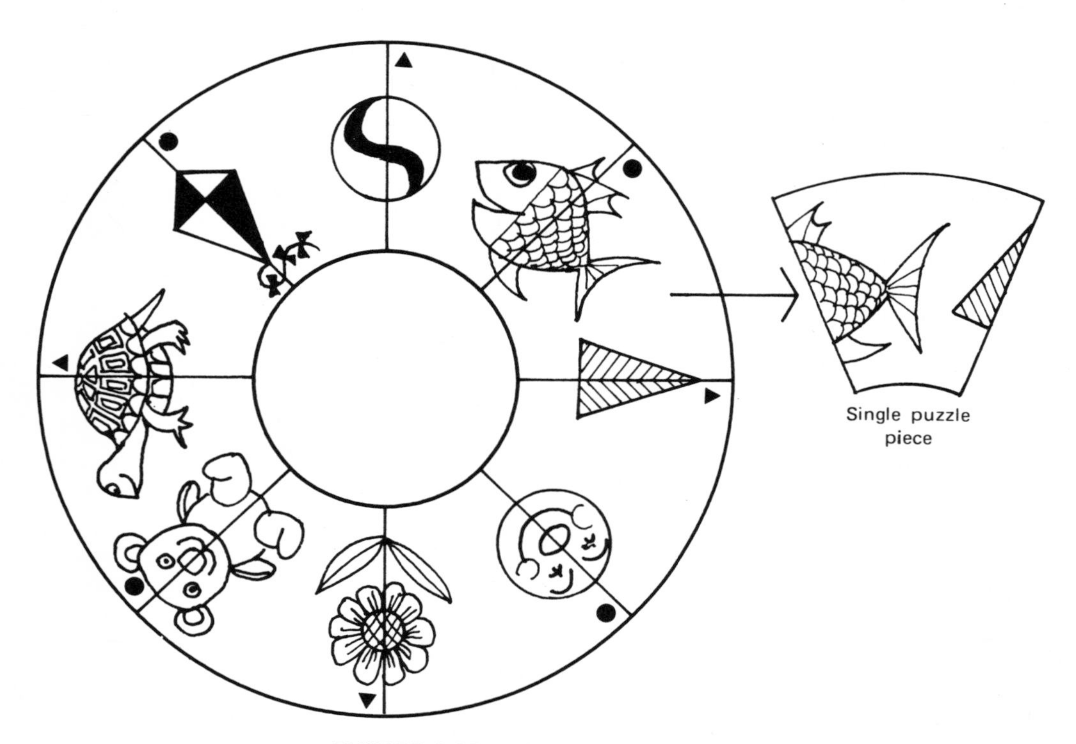

FIGURE 6.25. Sequence puzzle.

OBJECTIVES:

1. To develop visual sequencing abilities.
2. To develop form perception and figure-ground perception.
3. To help children learn how to take turns.

4. To help hyperactive children learn how to pace themselves.
5. To help children develop the concepts of visual association and closure.

ACTIVITIES:

1. Give all of the pieces of one code to one child and the pieces of the second code to another child. Have them put the puzzle together, as they take turns.
2. Have them predict the identity of an object on their puzzle pieces with limited information.

VISUALLY CODED MOVEMENT SEQUENCES

CONSTRUCTION:

Make visually coded movement sequences on task cards, as shown in Figure 6.26. Index cards may be utilized. Use a system of dots and dashes, long and short dashes, or musical notation for stimuli. Have blank cards in case you wish to make up some sequences as you work with the child. Allow the child to make his own sequence. In case you desire a set of permanent cards, it is a good suggestion to have them dry mounted for protection.

OBJECTIVES:

1. To develop visual sequential memory.
2. To help a child develop better powers of visual discrimination.
3. To develop better eye-foot and gross-motor coordination.
4. To enhance a child's power of visual decoding and manual encoding.

ACTIVITIES:

1. Look at a task card and verbally interpret what the card is telling you to perform.
2. While looking at the task card, perform the movements indicated.
3. Look at a task card and memorize the sequence. Put the card down and perform the movements from memory.
4. Look at the task card and then use your hands or fingers to clap or snap through the sequence. (— - - - — - long, short, short, short, long). Use other body parts to perform the same task.
5. Make up your own sequence on a card, memorize it and then, perform it.
6. Combine sequences so that you can memorize two, three, or four task card sequences at once and perform them without a break.

— long, walk, jump with both feet
– short, hop on one foot, run one step
• short, hop on one foot, run one step
○ slow walk
♩ step, walk normal, hop
♫ run, run
♩ ♫ • step-hop, skip, slide, gallop

FIGURE 6.26. Examples of visually coded movement sequences.

VISUALLY CODED MOVEMENT SEQUENCES

CONSTRUCTION:

A set of 3" x 5" index cards or a chalkboard is needed for this experience.

OBJECTIVES:

1. To assist children in developing the ability to use vision as a directing device for cognition and movement by introducing skills related to reading.
2. To help children learn basic locomotor skills, such as hopping, jumping, sliding, and rolling.
3. To teach children to move in response to visual cues.

ACTIVITIES:

1. Teach the child a set of directional action words. For example, run, jump, hop, skip, crawl, roll, draw, and cut are action words. Print these words on dividual 3 x 5" index cards and teach them to the child through a sight vocabulary method. If the child has trouble recognizing the words, draw a picture of the action next to the word so that the child can associate the word with the action.
2. Practice exposing a card and have the child execute the movement indicated. Say the word out loud. Then, encourage the child to move in response to only the visual cue.
3. Put individual sentences on a card or chalkboard that communicate specific movement directions for the child to execute.

Jump	3 x ↓	(Jump	3 times forward)
Roll	4 x →	(Roll	4 times to the right)
Draw	5 □	(Draw	5 squares)
Hop	5 x R	(Hop	5 times on the right foot)
Hop	4 x L	(Hop	4 times backwards on the left foot)

4. Read the word part to the child. He is to interpret the rest of the movement directions by himself. As he learns the directional words, he should figure out the entire sentence himself.
5. As the child develops performance in this routine of visually directed movement, the sentences should be enlarged. For example, Slide 3 x ← and Gallop 4 x ↓. From here he may be given three consecutive movement directions to carry out, again directed by the visualizing process.
6. Next, reverse the routine. Give the child the task of creating a directional sentence on an index card or chalkboard. The teacher or a partner should carry out the directions with the child deciding if the correct actions were executed. The instructions should occasionally be intentionally followed incorrectly in order to give the child the opportunity to discover that a mis-match has been made between his written instructions (a product of his visualization) and the subsequent movement pattern.

* The goal here is for the child to learn that the question being asked of him is "What does this tell us?" and not mere word recognition. He must learn that symbols convey *information* to him; that each communicates a specific kind of movement.

PLASTIC BOTTLES

CONSTRUCTION:

Gather several quart, half gallon, or gallon size plastic bottles. Decorate them, as shown in Figure 6.27. Animals are attractive to children and serve as motivational devices. Empty thread spools can be used for the legs. A face and tail can be devised with construction paper, pipe cleaners, felt paper, and the like. Cut a slit across the back of the bottle so that it runs sideways rather than lengthways. Make up sentence task cards or mathematics problems, as shown in the picture. Add task cards with activities related to exercises in position in space and

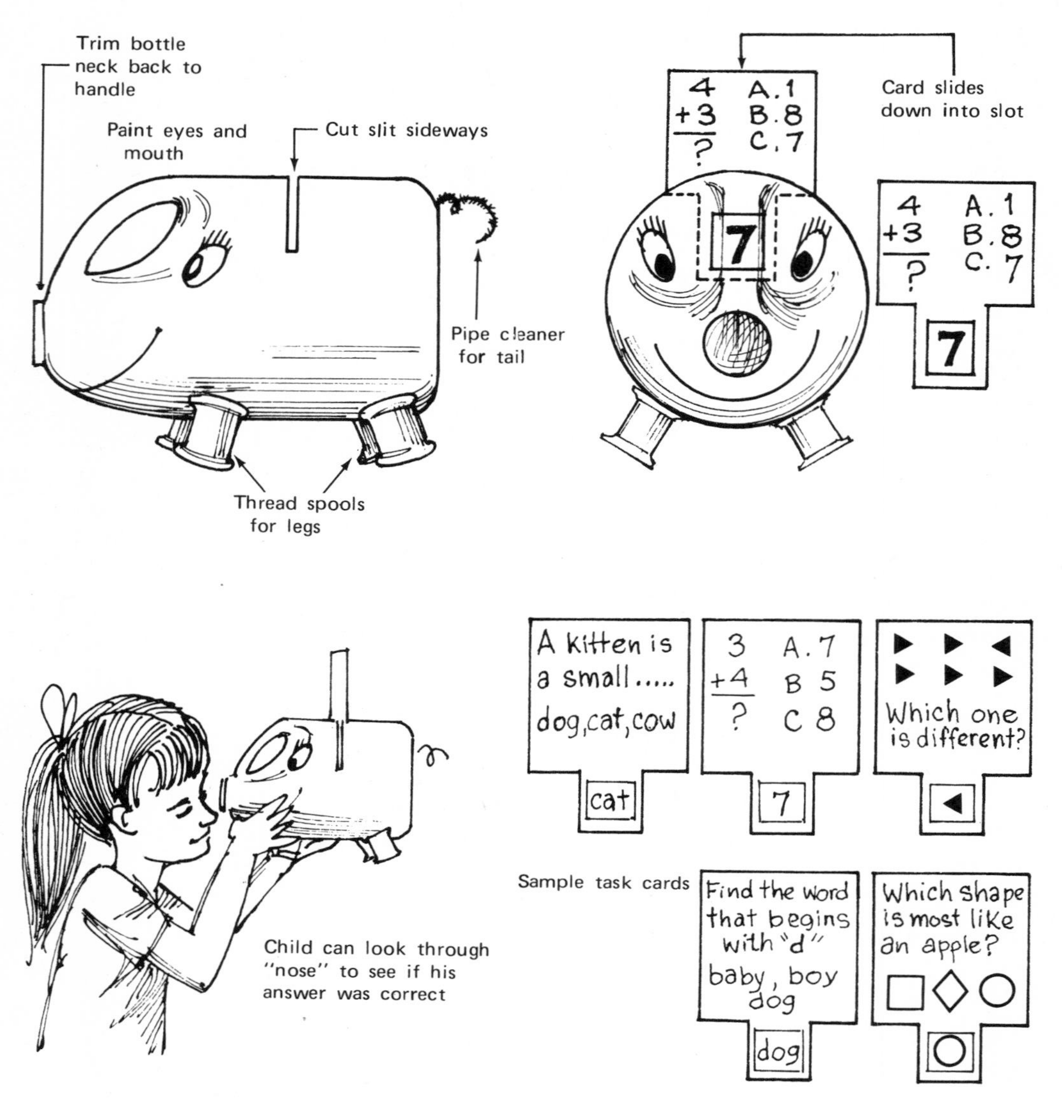

FIGURE 6.27. How to make a task card animal.

figure-ground perception. Make up a series of task cards according to the ability level of the children. Cut the bottom of each task card so that it may be inserted into the back of the animal. After the child chooses his answer, have him look through the nose to find the correct one. See Figure 6.27.

OBJECTIVES:

1. To develop visual decoding abilities.
2. To develop visual figure-ground perception.

3. To develop spatial awareness concepts concerning laterality, directionality, and position in space.
4. To enhance the child's ability to visually sort, search systematically, and process information.
5. To assist children in developing language and mathematical concepts.
6. To develop eye-hand coordination and fine-motor coordination.

ACTIVITIES:

1. Find the geometric shape or picture that is different from the rest of the objects.
2. Find the correct answer to the sentence or mathematics problem.
3. Find the words that begin with a "b" or a "d".

MIRROR TRACING

CONSTRUCTION:

A mirror tracer can be purchased from one of the companies listed in the resource section at the end of this book. The price will vary from $15.00 to $20.00. One can also be made inexpensively. A mirror, approximately 12 inch square, is needed along with two pieces of plywood with the same dimensions. One of the pieces of plywood should be used as the base and the second piece should be used as an eye shield. The mirror should be at a 90° angle, although it can be mounted so that it is adjustable. Make two braces to attach the shield to the brace, as seen in the picture. Use nuts and bolts with butterflies so that the shield can be adjusted at will. See Figure 6.28.

OBJECTIVES:

1. To utilize a visually distorted image to enhance the development of spatial awareness.
2. To develop eye-hand coordination while using a visually distorted image.
3. To develop tactile-kinesthetic perception.

ACTIVITIES:

1. Have the child grasp a pencil in his preferred hand and place his hand under the shield box. He should look into the mirror for his visual cues. The image will appear to be reversed and inverted. Use simple geometric forms, such as a circle,

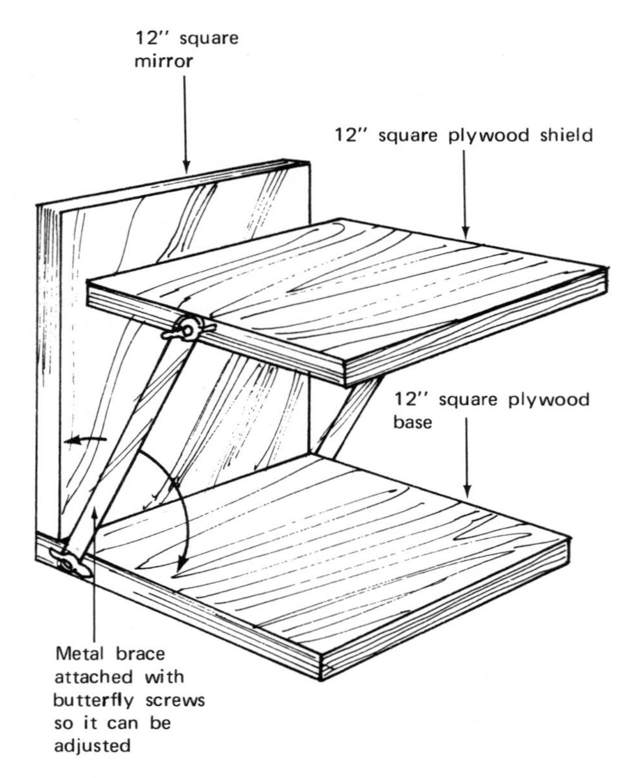

FIGURE 6.28. Mirror tracer.

triangle, square, star, or octagon. Forms should be double-lined and the child should try to stay between the lines.

2. Place a number of dots on a sheet of paper and place the paper on the base. Ask the child to connect the dots by moving in a clockwise or counterclockwise direction.
3. Ask the child to print his name or words from his spelling list while using the mirror tracer.
4. Ask the child to draw a human figure, house, or any other type of simple drawing, while using the mirror tracer.

* For further suggestions, use the mirror tracer for activities outlined in the Euclidean Geometry Games in the Spatial Awareness Chapter of this book (pp 62-63).

7
Auditory Perception

Auditory perception is one of the receptive phases of language development. It involves not only the ability to hear, but to also interpret and understand the message. For this reason, classroom teachers and physical educators should work closely with the speech and hearing therapist, for it is the latter who has specialized in corrective hearing therapy. There are, however, several concepts related to listening skills that can be developed by all teachers. These concepts are specific listening skills defined as auditory discrimination or decoding, auditory closure, auditory association, and auditory sequential memory (5, 26). Auditory discrimination or decoding is the ability to identify and accurately discriminate among sounds of different pitch, volume, and pattern. This includes the ability to distinguish one speech sound or noise from another. To help children develop this skill, activities should be designed which encourage them to compare the similarities and differences among speech sounds. Auditory closure is the tendency to complete an auditory sound or word, once initiated, and is related to the tendency of the human being to achieve completion in any behavior or mental act. Again, various activities that encourage children to develop skill in completing initiated words or sounds can be designed. Auditory association is the ability to relate spoken words in a meaningful way. As children learn synonyms or antonyms for words through association, they become more adept or flexible in using language to communicate. Auditory sequential memory is the ability to use recall to repeat correctly a sequence of auditory cues. This is a particularly important skill, as it relates directly to many school tasks, such as following directions, recalling the sequence of a story read or told by the teacher, or the sequencing of a rhythmical pattern.

AUDITORY DISCRIMINATION CANS

CONSTRUCTION:

Collect 10 to 20 small opaque containers, such as empty shoe polish cans or candy tins. Place different types of objects inside each of the cans which will make sounds when you shake them. Sand, paper clips, tacks, window screen, pebbles, rice, beans, and marbles are examples. Place the same substance in each of two cans. Put a code on the bottom of the cans so that the pairs may be identified.

OBJECTIVES:

1. To develop auditory discrimination.
2. To develop auditory sequential memory.
3. To develop auditory association.

ACTIVITIES:

1. Give the child the set of cans. Have him pair up the ones that sound alike.
2. Shake the can. Have the child find one that sounds the same. Have him find one that is louder or softer. Fine one that has a higher or lower pitch.
3. Change the numbers of objects in each of the cans. Have the child find the cans that sound similar, but have more or fewer objects in them.
4. Have the children try to guess what is in the cans. If they have problems, show them what is inside the cans and cover them again. Then, ask them what they think a particular object sounds like.
5. Shake two or three cans in a given sequence a specific number of times. Hand the cans to the child and ask him to repeat the sequence.

AUDITORY TAPES

CONSTRUCTION:

Reel-to-reel tapes or cassettes may be purchased commercially. The tapes should be used to record auditory stimuli that evoke auditory perception experiences in children.

OBJECTIVES:

1. To develop auditory figure-ground perception.
2. To develop auditory sequential memory.
3. To develop auditory discrimination.

ACTIVITIES:

1. If a child is distracted by various environmental stimuli, tape record directions of tasks you wish him to perform. Then, have him place earphones connected to the tape recorder over his ears to block out extraneous stimuli.
2. If a child learns best through auditory perception, record stories, mathematical problems, and other academic tasks for him.
3. Tape record sound effects from records or from the child's environment, such as sirens, auto horns, toilets flushing, water running, and passing trains. Teach the child to recognize each of the sounds.
4. Mask sounds by placing distracting noises in the background. See if he can distinguish the figure from the background. In addition to the sounds in number 3, mask words with distracting background noises. See if the child can distinguish the figure from the background.
5. Record two or three sounds simultaneously or sequentially. Then, play them as the child listens. Challenge him to identify all of the sounds.
6. Develop a rhythmical pattern on a tape, such as clap (hands), clap, snap (fingers), snap, tap (hands on knees), tap. Repeat this pattern several times. Have the child perform the exercise with the directions and sounds on the tape. Then, stop the tape and see if the child can continue the response.
7. Play several similar sounds with variations of high or low pitch, loud or soft sounds, and harsh or mellow sounds. Have the child describe each sound that is made or compare two sounds that he hears.
8. Use words from the Wepman Auditory Discrimination Test or others that sound alike. Group the words in pairs on the tape. Ask the child to tell you if the words are the same or different.
9. Perform the activity in number 8 with words that do or do not rhyme. Ask the child to tell you if the words do or do not rhyme.
10. Perform the activity in number 8 with prefixes and suffixes. Ask the child to tell you if the words begin or end with the same sounds.
11. Develop a series of Whistle Cues (see Whistle Cues in this section) on an auditory tape. Have him/her listen to the auditory sequence. Then, have him write out or perform the sequence. For example, long, long, short, short, short would be –, –, -, -, - on paper or jump (on two feet), jump, hop, hop, hop in a motoric sequence.

SNAP, CLAP, AND TAP

CONSTRUCTION:

A record player and a record with 4/4 meter music may be used to perform these activities to music.

OBJECTIVES:

1. To develop auditory and visual sequential memory.
2. To increase temporal and rhythmical awareness.
3. To develop eye-hand and gross-motor coordination.

ACTIVITIES:

1. Snap your fingers, clap your hands, and tap your knees in a specific sequence, such as snap, snap, tap, tap, clap, and clap. Have the child repeat the sequence after you. Make up others.
2. Have the child close his eyes or turn around and perform this same activity. This time the child has to rely on auditory perception alone in order to interpret the sequence and then perform it.
3. Have the children work together in pairs to develop their own sequences. Play music and have them keep time to the music.
4. Give a sequence verbally to a child and have him perform it motorically without any practice.
5. Perform a sequence motorically and have the child verbally tell you what sequence was executed.

LUMMI STICKS

CONSTRUCTION:

Cut the handles of discarded mops or brooms into 12 inch lengths. Paint or decorate the sticks as desired.

OBJECTIVES:

1. To develop auditory sequential memory.
2. To increase temporal or rhythmical awareness.
3. To develop eye-hand coordination.

ACTIVITIES:

1. Beat out a rhythm, as the child watches you. For example, long, long, long, short or hard, hard, hard, and soft. Then, have the child beat out the same rhythm with his stick.
2. Perform the same activity while the child has his back turned, so that all he can rely on is his auditory perception.
3. Have the child develop his own sequence of beats that he can repeat or teach to another child.
4. Play a record with 4/4 meter music and have the child tap the head of one or two sticks (one or both hands at the same time) to the floor to the beat of the music.
5. Tap the heal of one or both sticks to the floor to the beat of the music.

FIGURE 7.1. **With a partner, beat the sticks together.**

6. While sitting on the floor with your legs spread apart, tap the sticks between your legs, both to the right and to the left, one on each side; or cross arms and tap one on each side.
7. Hit the sticks together in the air to the rhythm of the music while changing the position of the sticks in relation to your body.
8. Flip one or both sticks into the air and catch them to the rhythm of the music.
9. With a partner, tap your sticks to the rhythm of the music, using any of the previous patterns (4–8). Try to create a symmetrical sequence of eight, sixteen, or twenty-four counts which you can repeat.
10. Tap your sticks together with a partner to the rhythm of the music. Tap right to right, left to left, right to left, left to right, or cross the sticks and tap both at the same time. See Figure 7.1.
11. With a partner, flip your sticks to each other to the rhythm of the music. Flip right to left, left to right, right to right, left to left, or both sticks at the same time.
12. Nonrhythmical tasks.
    a. Starting at the base of the sticks, have the child walk his fingers up the length of each stick, using only one hand. Then, walk his fingers back down to the base of the stick.
    b. Have the child balance the stick across his hand or another body part, as he performs locomotor and nonlocomotor movements. To increase the difficulty of this task, have the child balance the stick on it's end on different body parts.

SPELLING LISTS

CONSTRUCTION:

There is no equipment involved in this activity. Use words from the child's vocabulary. List words ranging from simple to complex. If desired, develop a workbook of each dhild's vocabulary used to perform various suggested activities.

OBJECTIVES:

1. To develop auditory and visual sequential memory.
2. To develop auditory and visual closure.
3. To increase auditory discrimination.
4. To develop auditory association.
5. To develop auditory and manual encoding.

ACTIVITIES:

1. Spell a word. Have the child repeat it. Spell the same word incompletely or incorrectly. Have the child complete the spelling of the word or correct the mistakes. Say the word and have the child spell it.
2. Say a sequence of three or four words. Have the child repeat them to you in the same order.
3. Show the child a sequence of three or four words on a chart. After removing or covering the chart, have the child auditorily repeat the words to you or write the words sequentially in his notebook.
4. Say a word to the child or show him one on a chart. Have him make up a spoken or written sentence with that word.
5. Begin a word verbally or on a chart and have the child complete the word.
6. Give the child several scrambled letters and have him make a complete word. For example, "o - v - l - e" unscrambled is "love." Do this experience auditorily or in writing.
7. Play other word games. Tell the child a word and have him think of one that rhymes, means the same, means the opposite, starts with the same letter, ends with the same letter, and so on.

GROCERY STORE

CONSTRUCTION:

There is no equipment needed for this game. However, if the child's auditory sequential memory is underdeveloped, the items he suggests during his turn can be used for visual reinforcement.

OBJECTIVES:

1. To develop auditory sequential memory.
2. To develop auditory and visual association.

ACTIVITIES:

1. Have several children sit in a small circle. As the game begins, each child will sequentially tell what they bought at the grocery store. Before a child can say what he bought at the store, he must repeat, in sequence, all of the other purchases stated by the previous children. For example, "I'm going to the grocery store to buy a box of cookies." "I'm going to the grocery store to buy a box of cookies and a gallon

of milk." The list continues until someone in the group can no longer remember all of the previously mentioned items. If the children exhibit difficulty, bring in the items or draw pictures of them. As the children state their purchases, have each one place his into the center of the circle. From this visual clue, the children will be better able to make the correct sequence.

2. The game can be varied by changing the circumstances or setting to stimulate the interests of the children. Going to a toy store, department store, beach, zoo, farm, and museum are some of these experiences.

JUMP ROPE

CONSTRUCTION:

Purchase 3/8 or 1/2 inch sash cord in 100 foot hanks from a local hardware store. Short jump ropes should be cut in 8 foot lengths or long enough so that the ends reach the child's armpits when he is standing in thy middle of the rope on the floor. Long jump ropes should be cut in 16 foot lengths. Tie the ends, tape the ends, or dip them in varnish to prevent the rope from unravelling.

OBJECTIVES:

1. To develop the childrens' agility, muscular coordination, and endurance.
2. To develop eye-hand and eye-foot coordination.
3. To develop rhythmical awareness and a better sense of timing in children.
4. To increase temporal awareness.
5. To develop creativity by using the ropes to form geometric shapes, letters, and numbers.

ACTIVITIES:

1. Short jump rope.
   a. Place the rope on the floor in a long line and walk along it in as many ways as possible, using one, two, three, and four body parts as a base of support.
   b. Form the rope into different letters of the alphabet by using different body parts, such as the feet, elbows, or hands.
   c. Jump or hop from one side of the rope to the other in forward, backwards, and sideways directions.
   d. Form the rope into different numbers by solving simple addition, subtraction, multiplication, or division problems posed by the teacher.
   e. Turn the rope forward and try differ-

ent methods of jumping–rebound jump, jump on one foot, alternate feet when jumping, jump in place, progress forward in a run, and so on.

f. Turn the rope backwards and try the latter methods of jumping.
g. Turn the rope to one side of the body or from one side to the other in figure eight fashion while jumping in place to the rhythm of the turn.
h. Experiment with changes in tempo while first jumping slow and then fast.

2. Long jump rope.
   a. With partners holding the rope at different levels, try to jump over, as in high jumping.
   b. With partners cradling the rope from side to side, try to jump it.
   c. With partners turning the rope, try rebound jumping. Turn the rope at various speeds, starting slow and getting faster.
   d. Try to go in the "front door" and "back door" while the rope is turning.
   e. Try to jump the long rope while you try to jump a short rope you are turning alone.
3. Blindfold jumping.
   a. Blindfold the child and have him listen to the tempo of the rope hitting the floor. Each time the rope hits the floor, he should be in the air. As a result, he needs to anticipate and time his jump to precede the rope hitting the floor. Have the child practice jumping outside the rope while it is turning. Finally, have the child try to jump the turning rope while he is blindfolded.

WHISTLE CUES

CONSTRUCTION:

Purchase some small toy whistles from a store.

OBJECTIVES:

1. To develop auditory discrimination.
2. To develop auditory sequential memory.
3. To develop manual encoding abilities.

ACTIVITIES:

1. Stand behind the child and whistle a sequence. Use code signals, such as long, short, long, and the like. Have the child repeat the sequence.
2. Use the whistle to develop a sequence. Give the child a pencil and paper and have him write it. For example, "–, -, -, –" is long, short, short, and long.
3. Whistle a sequence of longs and shorts and have the child develop a motoric response, such as hopping on one foot on the short cues and jumping on two feet on the long cues. Whistle the whole sequence first, then have the child interpret the sequence by jumping and hopping.
4. Whistle a sequence, have the child write it on a paper, as in number 2 above; then, have the child interpret the written sequence by jumping and hopping.
5. Establish a whistle pattern, such as long,

short, short, long . . . Do not complete the sequence. Ask the child what cue follows.

6. Blow varying qualities of sound, on the whistle, such as loud, soft, harsh, or mellow. Ask the child to discriminate between these sounds.
7. Develop an auditory tape based on the previous activities.

8

Tactile-Kinesthetic Perception

Tactile-kinesthetic perception relates to the reception of incoming stimuli directly affecting the body and the awareness of the body's position in space. Our bodies are equipped with a variety of external and internal receptor organs designed to sense light touch, touch pressure, pain, temperature, texture, and the position of our various body parts in space (4, 36). To help children become more aware of tactile-kinesthetic concepts, haptic experiences should be designed to assist them in comparing various multi-sensory activities related to gradations of temperature, color, texture, touch, size, shape, and the like. Children should also be encouraged to be more aware of bodily positions as they move through space. It is assumed that, as children become more efficient receivers and processors of incoming stimuli, they will be better able to understand their environment and, in turn, become more able to express themselves verbally or manually through movement.

FIGURE 8.1. Texture notebook.

HAPTIC EXPERIENCES

CONSTRUCTION:

Haptic experiences deal with children learning to discriminate between various textures, shapes, colors, weights, sizes, tastes, smells, and temperatures in our environment. Thus the teacher is encouraged to innovate different methods children can use to discover their world. For example, teachers can conduct smelling and tasting parties that introduce new odors and flavors to children. To give the reader more specific guidelines and suggestions for developing haptic experiences, an illustration involving texture experiences follows. The teacher can make a notebook containing various cloth textures. As seen in the picture, one half of the notebook should consist of the different materials. The second half of the notebook should contain words made from the materials describing the texture. See Figure 8.1.

A second type of texture experience can be gained by collecting various

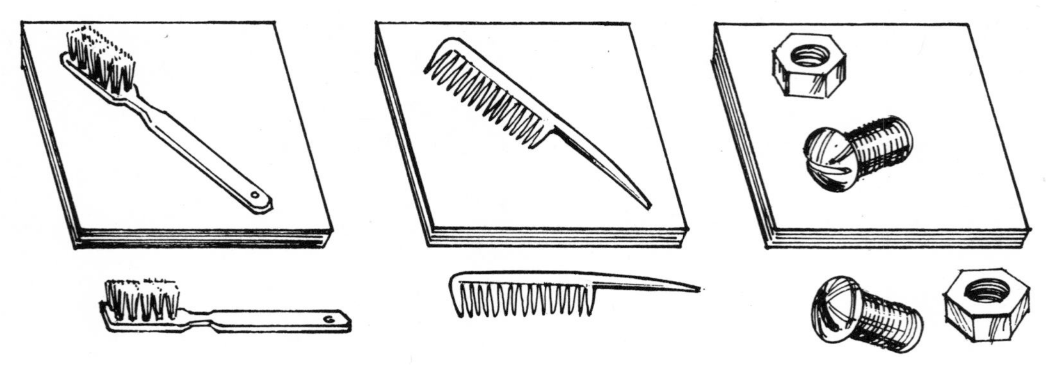

FIGURE 8.2. Texture boards.

textures and attaching them with staples to a large piece of cardboard or plywood. Tree bark, sponges, feathers, sandpaper, felt, silk, velvet, and similar textures from the child's environment are suggested. Keep a piece of each texture unattached so tha the child can feel it and find the one on the board that feels the same. See Figure 8.2.

OBJECTIVES:

1. To assist children in learning to discriminate among various textures, shapes, colors, weights, sizes, tastes, smells, and temperatures.
2. To develop tactile-kinesthetic awareness.
3. To develop fine-motor coordination.

ACTIVITIES:

1. When using the texture notebook, have the child identify the texture half of the book with the appropriate word describing the material.
2. When using the board, give the child a piece of the texturous material and have him identify the same material on the board.
3. Blindfold the child and give him an object. Have him find the one with the same texture on the board.
4. Blindfold the child and have him identify each of the objects on the board by naming it and describing it's texture.

CLAY AND FINGER PAINTS

CONSTRUCTION:

There are several good quality brands of clay and finger paint that are commercially made. However, the authors have found that children often learn as much from the process of making the clay or paint as they do when actually working with the materials. An interesting recipe for clay or dough follows:

> Mix one cup of peanut butter with 1¼ cups of dry milk, 1¼ cups of powdered sugar, and 1 cup of corn syrup. Mix and add chopped nuts, cocoanut, raisins, and so on, for a texture effect.

The children enjoy modeling the dough into various shapes. The dough is harmless, so if the children happen to get the dough under their finger nails and place their fingers into their mouth, there are no ill effects.

Children can mix their own finger paints by using water-based tempora powder. Add variability to the finger paints, such as flavor extracts, tint bases to change the colors, powder or oil to make the paint slippery or greasy, and textures, such as raisins or cocoanut for a lumpy texture. Although the children can't eat their products, the variability introduced to the paints will make the child more discriminatory in terms of color, smell, and texture.

OBJECTIVES:

1. To increase, to varying degrees, a child's haptic awareness of color, texture, smell, and taste.
2. To develop a child's tactile-kinesthetic perception.
3. To encourage a child to experiment with design, shape, dimension, flow, and so on.
4. To develop body awareness and self-concept.
5. To develop eye-hand coordination and fine-motor coordination.

ACTIVITIES:

1. Use the clay to model or sculpt the shapes of animal or human forms.
2. Flatten the clay on a board and use a stylus to draw pictures in the clay.
3. Mold the clay into the shapes of letters or numbers. Trace over the models with your fingers. What do the letters or numbers feel like?
4. Experiment with dimension by making a geometric design (circle, triangle, etc.). Then, using either clay or finger paints, make a similar one larger or smaller than the model.
5. Draw simple designs on a large sheet of paper with the finger paints. Emphasize smooth free-flowing lines in contrast to angular lines with abrupt changes of direction. Play smooth free-flowing string or symphony music in contrast to percussive music to add awareness to this experience. Have the children reproduce the design they have drawn by walking out the pattern on the floor.
6. While working with finger paints, draw letters and numbers on the paper. Can you draw a letter or number with your eyes closed? See Figure 8.3.
7. Paint a model of your body on a large

FIGURE 8.3. **Finger paints.**

piece of paper and identify each of your body parts.

8. By using finger paints, draw pictures of people, animals, houses, or your favorite scenes.

PUZZLES

CONSTRUCTION:

There is no special construction involved with developing this activity. Simply use the puzzles and parquetry blocks referred to in Chapters 4 and 5.

OBJECTIVES:

1. To develop tactile-kinesthetic awareness.
2. To develop form perception.
3. To develop small muscle and fine-motor coordination.
4. To develop spatial awareness concepts.
5. To develop visual sequential memory.

ACTIVITIES:

1. Blindfold the child and ask him to pick up a block or puzzle piece. Ask him to identify it by describing its shape, size, texture, and so on.
2. Construct a pattern of parquetry blocks for the child to observe. Then ask him to close his eyes and make the same pattern. Start with two pieces and increase the difficulty according to the child's ability. See Figure 8.4.

FIGURE 8.4. Reproduce a pattern with your eyes closed.

3. Construct a pattern of parquetry blocks under a cover for the child to feel. Then ask him to construct the same pattern with the extra blocks.
4. Perform this same activity, but ask the child to draw the pattern he felt with a pencil and paper.
5. Give the child verbal instructions and ask him to place blocks in a specific order with his eyes closed.
6. Select a relatively easy puzzle with irregular shapes. Ask the child to put the puzzle together with his eyes closed.

FEELY BOXES

CONSTRUCTION:

Small cardboard boxes from the shoe box size to 18 to 24 inches make good containers for tactile experiences. Wooden boxes can also be constructed. Make a hole in one side of the box for the child to insert his hand. A curtain can be made for the hole or window to prevent the child from seeing the objects inside. One end or the top of the box should be hinged to permit easy access to the inside of the box. The outside of the box can also be decorated to make it more attractive. See Figure 8.5.

OBJECTIVES:

1. To develop tactile-kinesthetic awareness.
2. To develop an awareness of form perception.
3. To develop fine-motor coordination.

FIGURE 8.5. Feely box.

ACTIVITIES:

1. Place various geometric shapes inside the box. Have the child identify one object at a time by placing his hand through the window and feeling the objects.
2. Name an object (square) and have the child find one inside the box.
3. Have the child reach into the box, find an object, name it, pull it out, and see if his answer was correct.
4. Show the child an object and have him find a similar one inside the box.
5. Perform these same activities with sandpaper letters and numbers.
6. Place objects common to the child's everyday environment inside the box (paper clip, scissors, pencil, apple, comb, fork, etc.). Perform the same previous activities.

ESTIMATION GAMES

CONSTRUCTION:

There is no construction involved with these activities. All that is needed is a chalkboard and colored chalk or several pieces of paper and pencils.

OBJECTIVES:

1. To assist children in learning how to discriminate distance, size, shape, and direction by utilizing the kinesthetic sense.
2. To develop fine-motor and small-muscle coordination.
3. To develop visual and tactile sequential memory.

ACTIVITIES:

1. Show the child a shape. Have him close his eyes and draw it from memory. See Figure 8.6.
2. Have the child close his eyes and draw a straight line on the paper. Have him open his eyes and examine the line he drew. Then, have him close his eyes and draw another line the same distance. If the child is familiar with the increments of distances, use the terms inch, foot, and so on. For example, "Draw a six inch line."
3. Have the child draw a simple geometric shape with his eyes closed. Have him open his eyes and examine the figure he drew. Then, close his eyes and draw another figure of the same shape with the same dimensions.
4. Draw a simple form on a chart using free-flowing or angular lines. After allowing the child to visually examine the drawing, have him close his eyes and draw the same form reproducing the same shape, size, direction, and orientation on the paper.

FIGURE 8.6. **Reproduce a shape or letter with your eyes closed.**

5. Have the child draw a human figure, picture of a house, or an animal on the chalkboard or paper while he has his eyes closed.

6. Ask the child to draw a simple geometric figure, such as a square with his eyes closed. After each line is drawn, ask him to take one step away from the board and then return to draw the next line. The eyes must be closed at all times.

LETTERS AND NUMBERS

CONSTRUCTION:

Make several sets of letters and numbers out of sandpaper, felt cloth, other cloth textures, and paper that has been glued and sprinkled with glitter. Draw an outline of the letters and numbers on the paper or cloth with a pencil and then cut them out with a scissors.

Be sure to make capital and small letters. Use different grades of sandpaper from very fine to coarse sheets. The letters and numbers should be from 2 to 4 inches tall. See Figure 8.7.

OBJECTIVES:

1. To develop tactile-kinesthetic awareness.
2. To help children learn to recognize letters and numbers.

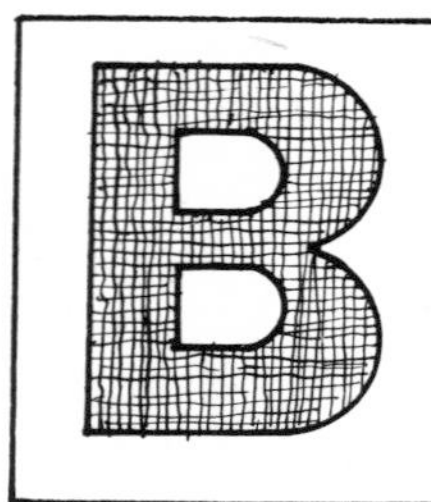
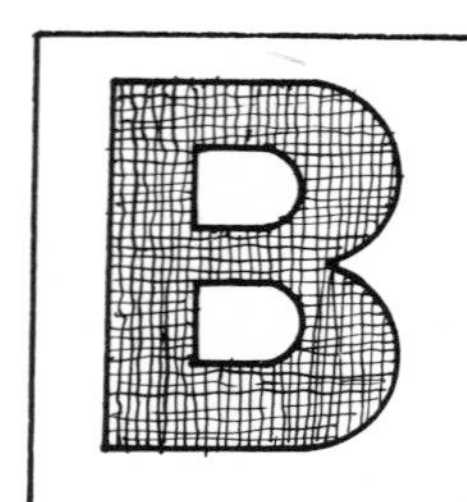

FIGURE 8.7. Textured letters and numbers.

3. To develop the spatial awareness concepts of laterality, directionality, position in space, and spatial relationships.

ACTIVITIES:

1. Trace over the letters or numbers with your eyes open. How do they feel? Notice how the different textures feel rough or smooth. Can you make the letters or numbers by drawing them in the air or on a piece of paper? Be aware of the direction in which you draw each letter or number.
2. Trace over one letter or number at a time with your eyes closed. Can you identify the letters or numbers when relying on your sense of touch? Again, notice the difference between textures.
3. Classify all of the letters or numbers. Find all of the "A's", "b's", "3's", "vowels", and so on.
4. Spell various words with the letters. How many different words can you make by changing only one letter in a word at a time? For example, lick, sick, dick, duck, suck, and luck.
5. Perform mathematical problems with the numbers. Try to add and subtract several numbers. Try to do some multiplication and division problems.

CHALKBOARDS

CONSTRUCTION:

Most schools have adequate chalkboards. Where necessity arises, an 8 x 4' sheet of masonite painted with chalkboard paint makes a good substitute. This type of board can provide hours of activity for the child when attached to a wall in his room or in the basement of his home. An easel can also be made if a portable board is desired. See Figure 8.8.

OBJECTIVES:

1. To develop eye-hand coordination and fine-motor coordination.
2. To encourage the development of form perception.
3. To help a child learn to cross the midline of his body.
4. To develop tactile-kinesthetic awareness and perception.

ACTIVITIES:

1. Allow the child to scribble and draw free-form designs.
2. Cover the whole board with chalk and have the child draw and scribble with his fingers.
3. Use templates to draw simple geometric

FIGURE 8.8. **Chalkboard activities.**

forms on the chalkboard. First, the child should use his fingers and then chalk to draw the forms. Begin with tracing activities, and progress to copying forms. Finally, have the child reproduce forms from memory.

4. Draw letters and numbers on the chalkboard either by copying a model or by drawing them from memory.
5. Single circle–Have the child press his nose against the board and make a large "X" on top of the mark made by his nose. Have the child draw a circle around the "X". The child should be taught to hold the chalk properly. The circle should be about 24 inches in diameter.
6. Double circles–Have the child hold a piece of chalk in each hand. Make a circle simultaneously with each hand.
   a. Draw the circles while moving the right hand clockwise and the left hand counterclockwise. Reverse the directions.
   b. Have the child move both hands clockwise, then counterclockwise.
7. Make horizontal lines with one or both hands at forehead, shoulder, and waist heights. Draw from right to left and vice versa. If the child has midline problems, have him stand on a raised box so that he can't move his feet.
8. Make vertical lines with one or both hands from forehead to waist height. Draw from top to bottom and vice versa.
9. Draw lazy eights on the chalkboard with one or both hands and change the direction of the drawing upon command.
10. Draw different motifs on the board and ask the child to reproduce a copy of the form. See Figure 8.9.
11. Draw a large clock on the board. Have the child place his hands on the numbers of the clock upon command. Use unilateral, bilateral, and contralateral movements.
12. Place dots on the board in sequence and have the child connect them. Start with simple sequences and work into situations where the child is required to cross the midline and intersect lines.

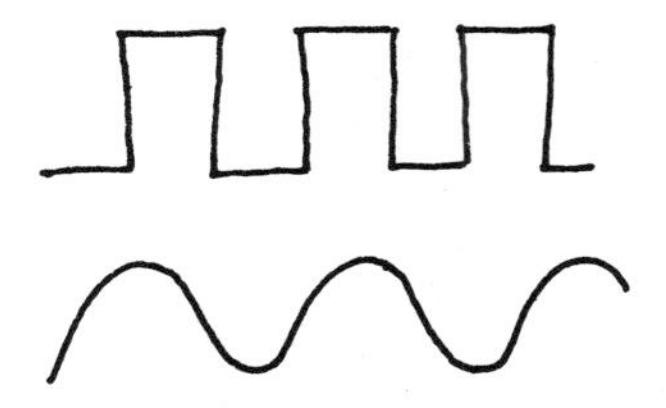

FIGURE 8.9. Motifs.

13. Rainbow writing–Have the child trace lines with colored chalk. Use different colors for each line to produce a rainbow effect.
14. Blind writing–Have the child dip his finger in a bowl of water and make letters and numbers with his wet finger while his eyes are closed.
15. Have the child draw pictures of people, animals, houses, toys, and other objects. Stress free-flowing natural movement of the arm, rather than restricted movement of the small muscles. This is indicated by placement of the wrist on the board.

MAZE WALKING

CONSTRUCTION:

Design a tunnel or a maze by connecting 4 x 6' folding gymnastic mats or large cardboard appliance boxes. Because many schools cannot afford to buy an adequate number of gymnastic mats, perhaps the cardboard boxes are a more pratical suggestion. They can be collected freely from hardware stores, department stores, and appliance stores. Lay the boxes on their sides and cut out the ends to make a tunnel. To make a corner, cut the proper dimensions out of the sides of two boxes, as seen in Figure 8.9. Insert one box into the other to make a fit. It is not necessary to attach one box to another. In fact it is easier to disassemble and rearrange if the boxes are not permanently attached. A maze can be made by standing the boxes on end. Cut holes in the boxes to serve as doors. They can be completely cut out or cut out on three sides and hinged. Doors can be cut out in different geometric shapes, such as circles, squares, triangles, and the like. Mazes should include a beginning, dead ends, and exit. The tops can be cut out of the boxes so that the teacher can see the child at all times. Boxes can be color-coded so that the child can learn to progress through the maze according to a color code. Boxes can also be painted for decoration. See Figures 8.10 and 8.11.

OBJECTIVES:

1. To assist in the development of spatial and kinesthetic awareness.

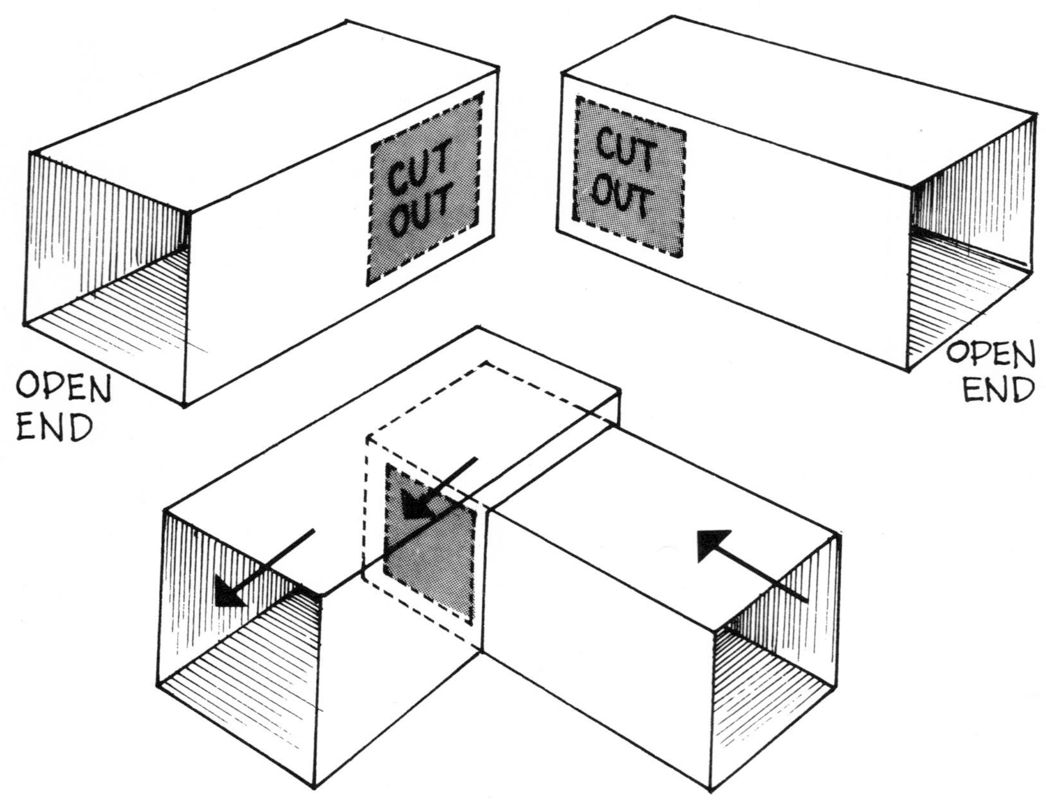

FIGURE 8.10. **Cardboard box tunnel.**

2. To develop visual, auditory, and tactile memory.
3. To teach children to give verbal directions (auditory encoding).

ACTIVITIES:

1. Have the child crawl or walk through the maze with his eyes open.
2. Lead with various body parts as you progress through the tunnel or maze. Lead with your head, feet, elbow, or knees.
3. Use a color code to progress through the tunnel or maze. When the child learns the correct sequence, he will be able to progress through the tunnel or maze from memory.
4. Have the child close his eyes or blindfold him. Have him progress through the tunnel from visual and tactile memory.
5. While blindfolded, have the child attempt to walk through a maze while relying on the tactile guidance of a partner to lead him through.
6. While blindfolded, have a child attempt to walk through a maze while relying on auditory directions from a partner.

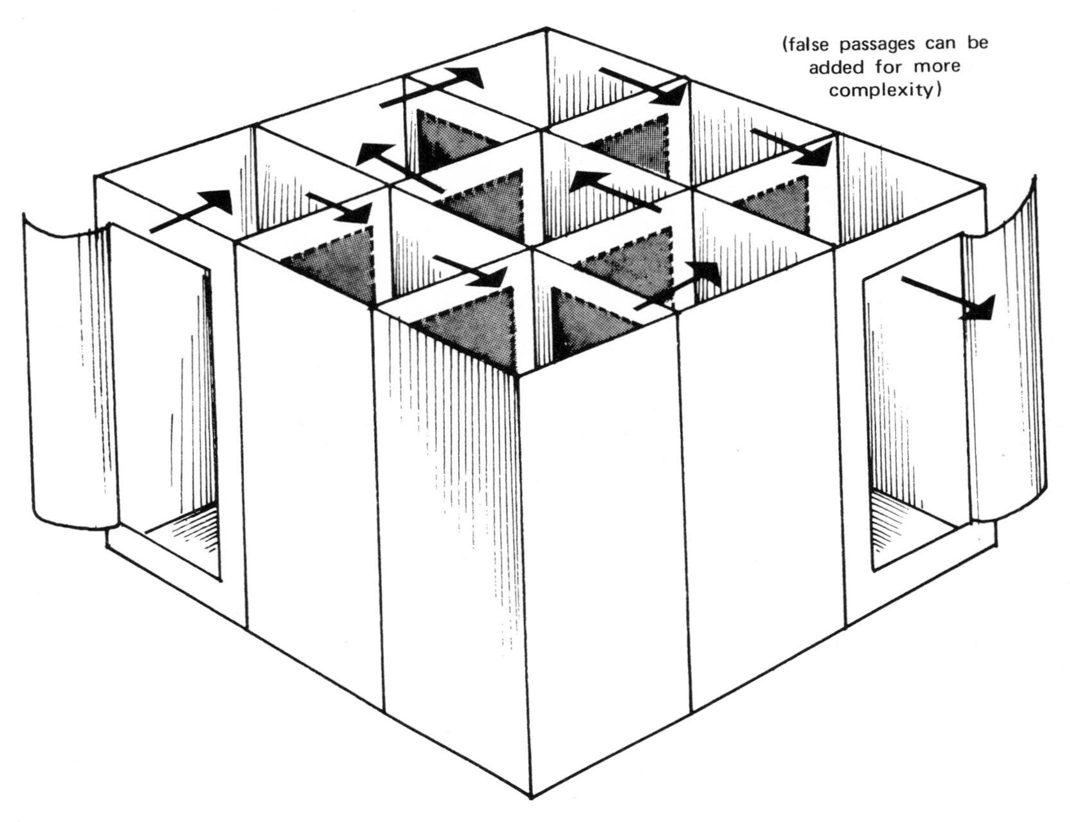

FIGURE 8.11. **Cardboard box maze.**

BINOCULARS

CONSTRUCTION:

A good set of binoculars can be purchased from an optics or sports department in a hardware or department store. Many times a discount will be allowed if you show the company that they will be used for school purposes. Perhaps the company will even donate a pair for educational purposes! No construction is needed.

OBJECTIVES:

1. To assist in the development of spatial awareness.
2. To develop balance.
3. To learn to move in space while being cued by a visually distorted image. See Figure 8.12.

FIGURE 8.12. Walk a straight line while looking through binoculars.

ACTIVITIES:

1. Walk on a line on the floor while looking through the large end of a pair of binoculars. See Figure 8.12.
2. Perform the same experience while looking through the small end of a pair of binoculars.
3. Walk through a maze while looking through the large or small end of a pair of binoculars.
4. Walk through an obstacle course while looking through the large or small end of a pair of binoculars.

BEANBAGS

CONSTRUCTION:

Use old scraps of material or felt and dried peas, navy beans, or styrofoam pellets from old beanbag chairs. Cut the material into the shapes of geometric forms, letters, or numbers. Sew two of the same forms together with strong thread until only a small hole remains. Fill the bag with beans and complete the sewing. See Figure 8.13.

OBJECTIVES:

1. To develop tactile-kinesthetic awareness concepts through manipulation and estimation of the distances which an object is thrown.

FIGURE 8.13. Letter, shape, and number beanbags.

2. To help children learn to distinguish shapes, numbers, and letters.
3. To develop fine-motor and eye-hand coordination.
4. To control an object and make it creative and functional.
5. To give children an opportunity to explore the relationship of an object to the surrounding area.

ACTIVITIES:

1. Tactile-kinesthetic awareness.
    a. Have the child throw at a target several times and count the number of times the target is hit.
    b. Throw the beanbag underhanded a given distance on the floor. Throw a second and third beanbag and try to make them land right on top of the first beanbag.
    c. Stand from 5 to 10 feet away from a target on the floor, such as a waste-basket. Throw several beanbags at the target and count the number that go into the wastebasket.
    d. After performing the above activity, blindfold the child and have him try to throw the bags into the waste-basket. Give the child verbal rein-forcement as to the placement of each bag so that throwing corrections can be made.
    e. Perform the same activity, only look through the large end of a pair of binoculars. If prism lenses are avail-able, use a pair which refract the images approximately 10°.
    f. Set up a number line measured in feet on the floor. Tell the child to state a distance and attempt to throw the beanbag that far.
2. Letter, shape, and number recognition.
    a. Throw the letter, shape, or number to the child and have him identify what it is, as they catch it.
    b. Have the children spell words with the letter beanbags. Then assume the shapes of the word's letters with their bodies.
    c. Give the children mathematics prob-lems involving addition, subtraction, multiplication, and division accord-ing to their ability level and have them solve the problems by using the beanbags. Then tell them to assume the shape of the numerical answer with their bodies.
3. Individual activities.
    a. Sit down, lie down, kneel, or close your eyes and do something with the beanbag.
    b. Place the beanbag on the floor and move around it or over it in several different ways.
    c. Move through general space while balancing the beanbag on your head. Can you change levels by standing up and stooping down?
    d. Balance the beanbag on another part of your body as you move through space.
    e. Throw the beanbag up in the air and catch it while you move about the room.
    f. Throw and catch the beanbag from one hand to another as you move about the room.
    g. Move the beanbag around your body as you change levels.
    h. Throw the beanbag into the air and use different parts of your body to catch it.
    i. Use various parts of your body to throw the beanbag into the air.
    j. Lie on your back and try to throw the beanbag into the air and catch it.
    k. Place the beanbag on the floor and build different types of bridges over it with your body. Make bridges with three, four, five, or six body parts touching the floor.
    l. Place the beanbag on the floor and move it with different body parts. Try your elbow, knee, heel, or head.
4. Partner activities.
    a. Find a partner and exchange bean-bags on the floor.
    b. Find several different ways to ex-change beanbags in the air.

c. Exchange beanbags with a partner, using body parts other than your hands.
d. Move about the room with a partner and exchange beanbags.

* For additional ideas for beanbag uses, the reader is encouraged to try some of the activities suggested for ball uses in another section of this book (pp 107-111).

Eye-Hand, Eye-Foot, and Fine-Motor Coordination

Eye-hand, eye-foot, and fine-motor coordination are means of manual expression (6). Each skill relates to how we communicate nonverbally. Each skill refers to a perceptual-motor match where the eyes and hands or feet act in a unified or coordinated fashion to produce an efficient movement (27). The development of eye-hand coordination will enable a child to execute gross-motor acts efficiently, thereby enabling him to perform the various movement patterns essential in work and play. Eye-foot coordination is essential in the development of gross-motor patterns relating to locomotor, rhythmical, and manipulative tasks. Fine-motor coordination refers directly to the development of the small muscles of the hand which are used in tasks, such as writing, spinning tops, or assembling nuts and bolts. It is assumed that successful execution of all of these manual tasks will also lead toward the development of a positive self-concept and academic success.

FIGURE 9.1. Children can throw objects through holes in the target.

FORM TARGETS

CONSTRUCTION:

Targets with simple geometric forms cut out can be made from pieces of plywood or cardboard boxes. If possible back supports should be made to allow the target to be placed at various angles toward which the child can aim. The targets can be painted to create a more complete and attractive project (see Figure 9.1).

OBJECTIVES:

1. To help children develop eye-hand and gross-motor coordination.
2. To help children develop better concepts of form perception.
3. To develop a child's tactile-kinesthetic perception.
4. To assist the child in learning how to plan motor activities.

ACTIVITIES:

1. Have the child stand at various distances from the target and try to throw objects, such as beanbags, through the holes.
2. Vary the angle of the target from perpendicular to the floor to parallel with the floor and have the child try to toss objects through the holes.
3. Tell the child to toss objects through specific holes in the target by aiming at specific geometric forms on command.
4. Have the child trace his fingers around the openings in the target to identify various forms.

IMPROVISED RACKETS

CONSTRUCTION:

One metal coat hanger, a pair of nylon hose, and some masking tape or adhesive tape are all that is needed to construct each racket. Bend the hanger into a diamond shape and straighten the hanger hook. Insert the hanger into the hose, making sure to push the end of the hanger snugly into the toe of the stocking. Pull the stocking tightly around the hanger and gather the loose ends by twisting them around the handle. Tape the handle at the base of the diamond. Repeat this same procedure for a second nylon hose leg. Then bend half of the handle back toward the base of the diamond and tape the whole handle. See Figure 9.2.

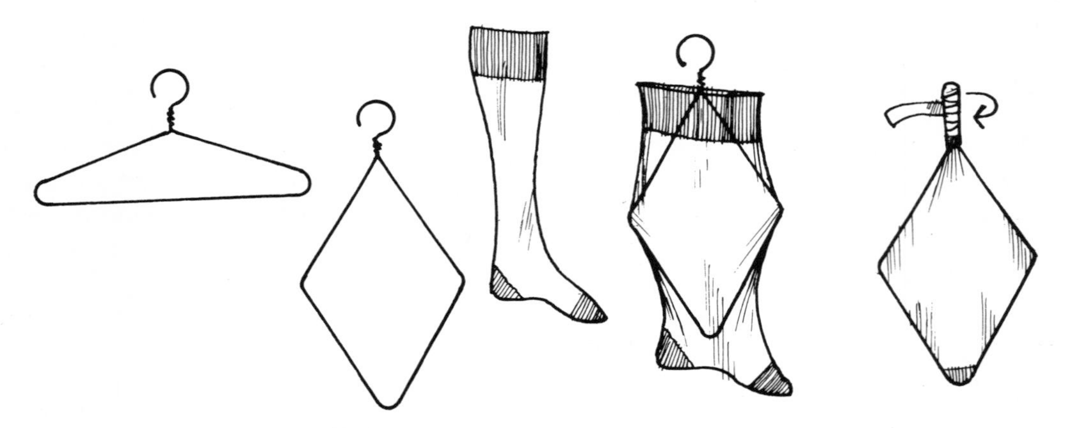

FIGURE 9.2. **Nylon hose rackets.**

OBJECTIVES:

1. To develop eye-hand coordination.
2. To develop better striking abilities.
3. To increase a child's spatial awareness.
4. To teach teamwork and cooperation through partner and team activities.

ACTIVITIES:

(small and/or lightweight balls, such as ping pong balls, plastic golf balls, yarn balls, or badminton birds work best with this type of racket)

1. Balance the ball on your racket in forehand and backhand positions and move about the room.
2. Volley the ball into the air continuously while using either the right or left hand in forehand or backhand position. Change hands and change hand positions.
3. Hit the ball into the air as high as you can and control it.
4. Volley the ball against the wall as many times as you can.
5. Volley the ball back and forth with a partner.
6. As the ball comes to you, see if you can stop its momentum and make it come to rest on your racket.
7. Try hitting the ball in different ways—over your head, behind your back, under your legs, and so on.
8. Given a racket, a ball, and some other pieces of equipment, such as a rope or hoop, create a new game with your equipment.
9. Play a modified game of badminton or tennis with your rackets.

BATTING TEE

CONSTRUCTION:

Batting tees can be made in different ways. One method is to cut the top end of a wiffle ball bat and then slip the bat through the end of a traffic cone, as seen in Figure 9.2. To make a wooden tee, obtain a piece of 2' x 2' x 3/4" plywood to use as a base. Anchor a 1½ to 2" pole or dowel to the center of the base with nails or screws. The pole should be 2 to 3 feet tall, depending on the size of the children. Attach a piece of hose, such as a radiator water hose, with an adjustable clamp to the pole. The hose can then be adjusted to the height of the children. Place the ball on top of the hose for the children to hit. The flexibility in the hose keeps the children from being jarred when they hit the ball. See Figure 9.3.

FIGURE 9.3. Batting tee.

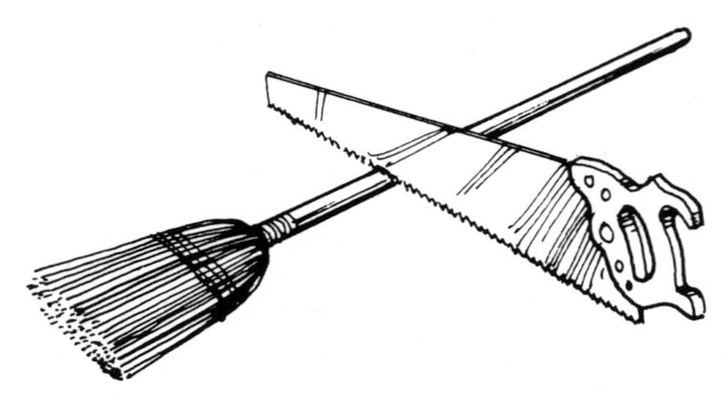

FIGURE 9.4. Saw off the handles of discarded broom and map handles.

OBJECTIVES:

1. To develop eye-hand coordination.
2. To develop concepts of laterality and directionality.
3. To develop gross-motor coordination.

ACTIVITIES:

1. Use a table tennis racket, paddle ball racket, or wiffle bat and have the child hit a wiffle ball, yarn ball, or Nerf (sponge) ball off the tee. As the child's skill increases, have him hit grounders, fly balls, or place his hits to right or left field.
2. Use a softball bat and softball to perform the same activities.
3. Have a group of children play "Tee Ball." No pitcher is required. The rules are similar to softball or can be modified for any situation.

BROOM STICKS AND MOP HANDLES

CONSTRUCTION:

The handles of broken or worn-out brooms and mops may be collected from home, school, or industry and used for a variety of purposes in perceptual-motor development. For various striking activities, leave the broom intact and use the head as the striking surface. To construct a wand, cut the head off the broom or mop so that it is 3 to 4 feet long. See Figure 9.4.

OBJECTIVES:

1. To develop eye-hand and eye-foot coordination.

2. To develop various aspects of physical fitness, such as balance, flexibility, agility, and strength.
3. To help children develop spatial awareness.
4. To develop rhythmical and/or temporal concepts.

ACTIVITIES:

1. Whole brooms.
   a. Ride the broom like a horse.
   b. Play various hockey games by using the broom as a stick and a wiffle ball, shaving can top, or similar object as a puck.
   c. Use the broom as a bat for stick ball.
2. Wands.
   a. Balance the wand on your hand, head, or another body part in a horizontal position as you move about the room.
   b. Balance the wand in a vertical position on various body parts.
   c. Balance the wand on the floor in a vertical position. Release grasp, turn around 360°, and regrasp before the wand falls to the floor.
   d. Stand a wand on one end and hold on to it. Bring one of your feet over the stick, letting go of it and catching it again before it falls to the ground.
   e. With a partner, balance your wand on the floor in a vertical position. Release the grasp of your wand and catch your partner's wand before it falls to the ground.
   f. Roll the wand on the floor and jump over it.
   g. Place several wands on the floor at varying distances apart and hop or jump over the wands while varying your direction, level, speed, or force of movement.
   h. Hold onto the wand at each end and jump through the arms and over the wand without releasing your grip. See Figure 9.5.
   i. Have a partner rotate a wand around himself at a level near the floor. Jump the wand as it passes under your feet.
   j. Grasp the wand with your arms apart. Keeping your arms straight, pass the wand over your head and down to the back.
   k. Grasp the wand with both hands, palms facing away from your body. Step over the wand with one leg at a time. Bring the wand over the head. Step around one arm with the leg on the same side. Place the leg that went around the arm between you and the wand. Pull on the wand until it is all the way over your back and head. Finally, step over the wand until you are back to the original position in which you started.
   l. Hold the wand in a vertical position with the end of the wand on the floor. Twist so that you pass under the arm holding the wand without letting go of it, taking it off the floor, or touching your knee to the floor.
   m. Throw the wand as far as you can, as if you were throwing a javelin.
   n. Twirl the wand, as if it were a baton.
   o. Walk on the wand placed on the floor, as if it were a tightrope.

FIGURE 9.5. **Jump over stick.**

p. Perform a cartwheel or other leaping, hurdling, and jumping stunts over the wand.
q. With a partner sitting or standing on a carpet sample or gym scooter and holding on to a wand, pull him around the floor.
r. Try to perform limbo stunts by passing under a wand held at levels closer and closer to the floor.
s. Have partners play tug-of-war while grasping the wand in several different positions. For example, try holding the wand while you are face to face, back to back, or place the wand between your legs.
t. Try wrestling a wand away from your partner. The wand should be gripped so that the children's hands are alternately placed. On command, each child tries to make the other child let go of the wand by twisting the wand back and forth and applying pressure.
u. With a partner, hold onto the wand as in the same activity as above. On command, each child tries to touch the right end (his right side) of the wand to the ground. Whoever touches his end to the ground is the winner.
v. With each partner holding onto a wand and crossing the wands in the middle, try to push your opponent off balance or across a line that is drawn behind each person (about three feet).
w. Divide the class into relay teams. Have them run a designated distance, place the wand on the floor with the palm of one hand on top of the wand, and rotate around the wand three times (Dizzy Izzy). Each person takes his turn.
x. Perform other relays by balancing the wand on various body parts, carrying the wand, or using the wand to strike objects on the floor through an obstacle course.
y. Attach a weight to the wand on the end of a rope. Try to roll the weight up to the wand by using forward and backward rolling motions.
z. Tie a rope to the end of the wand. Attach a brick to the end of the rope. Have the child use the wand in this fashion as a counter-balancing agent, as he crosses a balance beam.

FIGURE 9.6. Attaching a brick to a long stick forces a child to balance while using both sides of his body.

This is good for children who tend to isolate one side of their body by holding one arm close to the trunk. See Figure 9.6.

aa. Use the broomstick or mop handle as a baton and perform various simple twirling activities. First, use one hand, then the other.

ab. Use the wand to form different letters of the alphabet. Get into small groups and spell words. See Figure 9.7.

HOOPS

CONSTRUCTION:

Commercial hoops are available in several sizes and colors. However, they are not as well constructed as those that can be made from plastic plumbing tubing or garden hose. Also, the cost of making a hoop is only half as much as the cost of buying one. Homemade hoops are more rigid, can be made without the use of staples, and the circumference can be made to meet any specific measurement. The plastic tubing, garden hose, and couplers for joining ends can be purchased at most hardware stores. Tubing, garden hose, and couplers are available in various sizes from ½ to 2 inch thicknesses. They are priced by the foot and according to

FIGURE 9.7. Body alphabet.

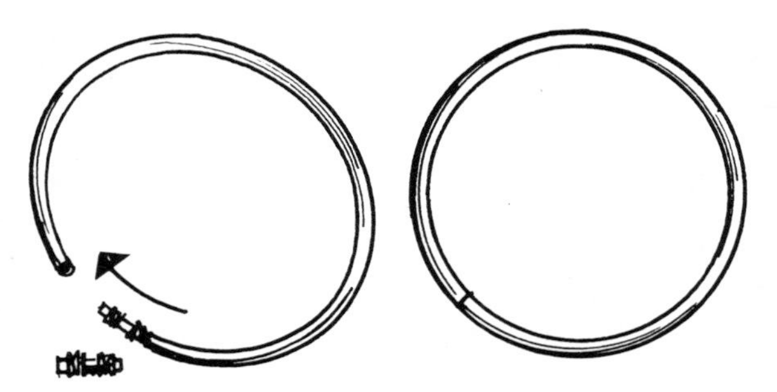

FIGURE 9.8. Hoops.

size. Use garden hose to make small hoops 6 to 12 inches in diameter and plumbing tubing to make large hoops 3 to 8 feet in diameter. Cut the ends of the hose or tubing so that they will meet and join smoothly. Heat both ends of the hose or tubing by dipping them in hot water. This enables it to expand to allow the coupler to fit into the ends. Insert the coupler into the hose or tubing and allow to cool. See Figure 9.8.

OBJECTIVES:

1. To develop a child's spatial awareness.
2. To increase gross-motor coordination.
3. To develop eye-hand and eye-foot coordination.
4. To develop body rhythm and an awareness of tempo.

ACTIVITIES:

1. Hoops are made for rolling.
   a. Roll the hoop, run ahead of it, catch it, and roll it back.
   b. Roll the hoop, then run and hop, dive, or jump through it as it rolls along on the floor.
   c. Roll the hoop or spin the hoop in place and see how many times you can jump in and out of it before it stops.
   d. How long can you keep the hoop rolling? How fast can you roll it? How slowly can you roll it? Can you roll the hoop and make it come back to you?
   e. Can you roll the hoop with some part of your body besides your hands? Can you keep the hoop rolling by using a stick?
   f. Can you bounce a ball through the hoop while it is rolling? How many times can you pass a ball back and forth to your partner before the hoop stops rolling?
2. Hoops are made for throwing and catching.
   a. Toss the hoop into the air and catch it.
   b. Toss the hoop into the air so that it comes down horizontally with you in the center.
   c. Toss the hoop from a low position.

Jump to a high position and catch it.

d. Toss the hoop up high in the air and turn around before you catch it.
e. Using two or more hoops, try to juggle them.
f. Use a plastic bottle with some sand in the bottom for stability, as a target on the floor. Throw small hoops from various distances and try to ring the target.
g. Play a game of Deck Tennis with some small hoops.

3. Hoops are made for rotating.
   a. Rotate the hoop around various body parts—wrists, arms, neck, legs, ankles, hips, or waist. Can you keep more than one hoop going at a time?
   b. Can you walk, run, skip, hop, or gallop while you keep the hoop rotating around you?
   c. Can you walk across a balance beam while rotating the hoop around you?
   d. Can you keep the hoop rotating while your body assumes different poses?
   e. Can you bounce a ball or jump while you keep the hoop rotating around you?
4. Hoops are made for self-testing and gymnastics activities.
   a. Can you jump the hoop as if you are jumping rope?
   b. Can you start the hoop at your neck, twirl the hoop down to your feet and back up to your neck?
   c. Can you do a cartwheel by placing your hands in the middle of the hoop?
   d. Can you do a handspring by placing your hands in the middle of the hoop?
   e. Crawl through the hoop using various body positions, levels, and bases of support.
   f. Do tumbling stunts with hoop and partner. For example, perform a dive roll through a vertical hoop held by a partner.
   g. Use one small hoop with a partner. Both children kneel down on the floor and grasp the hoop with one hand. Try to wrestle the hoop away from your partner. The first one who gains sole possession of the hoop wins the match.
   h. Work with a partner with each holding a small hoop. Kneel down on the floor. On the command, "Go," try to put your hoop around one of your partner's legs while trying to avoid the placement of his hoop around your leg.
5. Hoops are made for rhythmical and game activities.
   a. Any of the rolling, throwing, tossing, or rotating activites lend themselves to a variety of relay games and races.
   b. Place twenty to forty hoops on the floor in a pattern. This makes an excellent maze for running, skipping, hopping, jumping, and so on.
   c. Hoops can be used as targets for balls, beanbags, balloons, and the like. They provide good stationary targets or moving targets as well.
   d. One or two children can hold a hoop horizontally while another jumps over and under or in and out. Also try to perform cartwheels through held hoops.
   e. Given one ball and one hoop, make up a game to play with one or two partners.
   f. Any music with a definite beat can be used for creating rhythmic movements. Records are available with instructions for routines and exercises. What can you do with your hoop to keep in time with the beat of the music?

FIGURE 9.9. Perform exercises with inner tubes.

BICYCLE INNER TUBES

CONSTRUCTION:

Used or worn bicycle inner tubes may be obtained from bicycle repair shops for the asking. Children may also bring them from home as bicycle tires and tubes are replaced. Use adhesive tape to tape the valve stem to the tube.

OBJECTIVES:

1. To help children develop flexibility.
2. To help children develop muscles through isometric and isotonic excercises.
3. To help children develop body awareness (body parts).
4. To help children develop social interaction.
5. To help children have fun and enjoyment.

ACTIVITIES:

1. Individual Activities (should be held for 6 counts). See Figure 9.9.
   a. Stand with both feet on the inner tube; pull up with both arms (palms down).
   b. Same as item one, but reverse grip (palms up).
   c. Stand with both feet on the tube;

loop the tube behind the neck; from a crouch position, force the tube up with the neck.

d. Same as item three, but push arms out to the side.
e. Double the tube; loop the tube around one foot; balance on one foot; pull to the *side* with the other foot.
f. Repeat item five, using the other foot.
g. Double the tube; stand on the tube with one foot; pull the tube *up* with the other foot.
h. Repeat item seven, using the other foot.
i. Double the tube; hook behind one heel; keep other leg straight; point toe and lift leg straight forward; hold 6 counts.
j. Repeat item nine, using the other foot.
k. Double the tube; hold high over the head with straight arms; pull for six counts.
l. Double the tube; hold in front of body with arms straight; pull for six counts.
m. Using two doubled tubes, place one under each foot; pull up with hands and walk with stiff legs.

2. Partners.
   a. Sit with feet flat against partners; hook tube over feet; keep legs straight and use rowing technique; lie down, sit up, lie down, sit up, and so on.
   b. With tube around waist of both partners, have each partner take four steps out, four steps in.
   c. Make up a creative movement with your partner.
   d. What other partner stunts can be done;
3. Groups of Four.
   a. With tube held in hands, two partners stay in place while two back out; continue to work in and out, in and out, and so on.
   b. *All* walk in and out.
   c. All hold tube above heads; turn inside out.
   d. Travel in a circle; walk, skip, run, hop, and so on.
   e. Sit with legs straight, feet together; one and three lie down, two and four sit up; alternate up and down.

* Try other combinations with small group work.

* Make up new ideas and have the students create new activities.

Note: Be sure to *tape* down the valve of the inner tube so that it does not protrude and become a safety hazard.

PLASTIC BOTTLES

CONSTRUCTION:

Pint, quart, half-gallon and gallon milk containers, bleach bottles, soap dispensers, and so on, can be obtained from the home. Because they can be used in so many ways and the construction is so simple, only a brief description of the process will be mentioned. See Figure 9.10.

1. Boundary markers, goal markers, or obstacle markers can be made by pouring about two cups of sand into the bottle (so that the wind won't blow it over) and then painting with various numbers or symbols for the different activities. Keep the caps on the jugs.

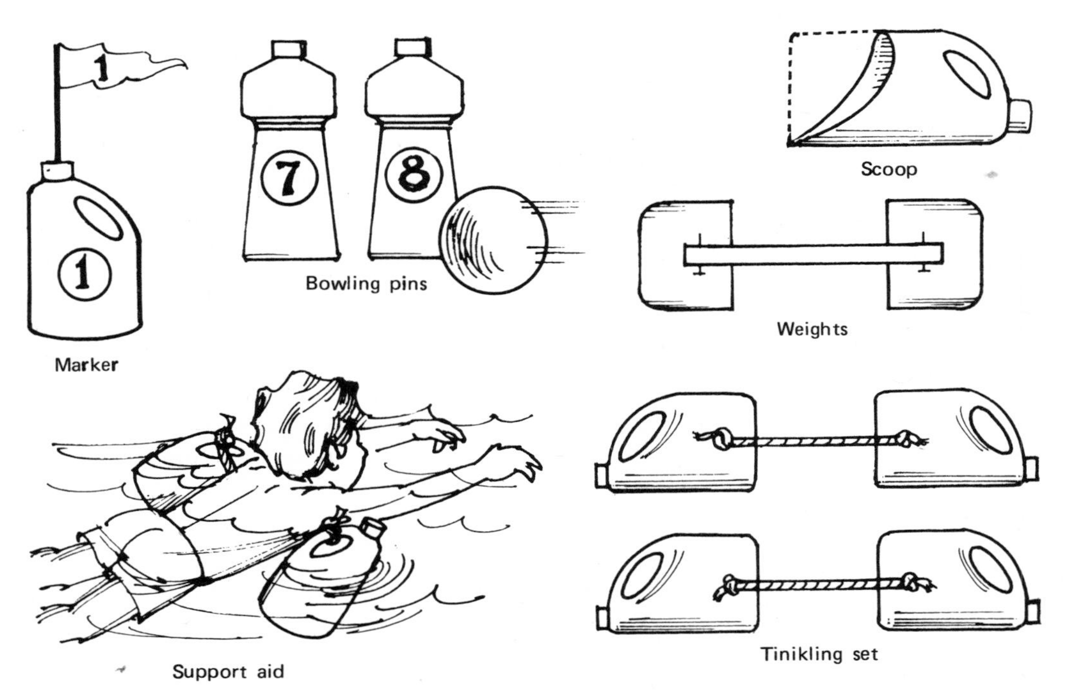

FIGURE 9.10. Plastic bottles have many uses.

2. Plastic bottles can be used as bowling pins. The jugs should be painted and may be stood on the bottom or cap, making it easier for small children to knock them over. Any size jug can be used.
3. Homemade weights can be made by filling the plastic jugs with cement and joining them with aluminum or galvanized lengths of pipe. Drill a 1/4 hole near the end of the pipe and insert a large nail through the hole. This will prevent the cement from pulling off the pipe. Pour the cement into one of the cans and place the pipe into the mixed cement making sure the pipe is placed in the middle and perpendicular to the bottom of the can. Let this one dry before repeating the process to the other end of the pipe.
4. Plastic bottles may also be used as support aids for beginning swimmers. They may be used individually or joined together with short pieces of rope.
5. A Tinikling (Phillipine folk dance) set can be made from four plastic jugs (one gallon size) and two pieces of rope of any desired length. Drill a hole in the center of the bottom of each jug slightly larger than the diameter of the rope. Next, insert the rope in the hole and feed it through until it comes out the mouth of the jug. Then tie a large knot in the end and slip it back into the jug. Repeat the procedure with the other end of the rope on another jug. Repeat the procedure with the other end of the rope on another jug. Repeat the procedure with the second rope and remaining jugs.
6. Scoops can be made by cutting the end and/or any part of the side out of the plastic bottles. Place tape around the edges which have been cut for protection. Paint the scoops as desired. All scoops should have the grip-type handles.

OBJECTIVES:

1. To develop eye-hand and eye-foot coordination.
2. To increase spatial awareness, laterality, and directionality.
3. To develop rhythmical awareness.
4. To increase gross-motor coordination.

ACTIVITIES:

(Because the plastic bottles are used primarily for markers and for lifting or supporting activities in the first four items in construction above, suggested activities will be limited to uses of the Tinikling sets and scoops.)

1. Tinikling is a folk dance originating from the Philippine Islands. An album containing Tinikling music is RCA Victor-LPM1619. Any record with 3/4 time may be substituted. Originally done with two people striking bamboo poles together at ankle level in an "in, out, out" rhythm pattern, the dance may be executed quite well with the innovative equipment. While the ropes are together, the participants jump with their feet on the outside of the ropes. While the ropes are apart for two counts, the participants jump with their feet on the inside of the ropes. To create a definite rhythmical beat, the plastic bottles should be struck together on the "in" beat and should be struck on the floor about shoulder width apart on the two "out" beats. Some possible patterns follow:
   a. Straddle the ropes with both feet on "in." Jump two times with feet on the inside on the two "out" beats.
   b. Do this same procedure, but turn a half turn as you jump while on the inside of the ropes.
   c. Start on the left side of the ropes. On the "in" beat, remain on the outside and place your weight on your left foot. On the two "out" beats, jump on the inside with your right foot two times. Continue the pattern—left, right, right, and so on.
   d. Do the same as above, only on the right side of the ropes. The pattern will be right, left, left, and the like.
   e. Start on the left side of the ropes and transfer to the right side. On the "in" beat place your weight on your left foot. On the first "out" beat, jump to the inside with your right foot. on the second "out" beat, jump on the inside with your left foot. As the ropes come together again jump to the outside and land on your right foot. Continue back into the center with the left and right on the next measure. Continue the crossing pattern, crossing from one side of the ropes to the other on each new measure.
   f. Think up new patterns and try them on your own!
2. Plastic bottles with the bottoms cut out can be used for many throwing and catching activities. Various sized balls may be used.
   a. Throw the ball up to yourself and catch it with your scoop.
   b. Throw the ball against a wall and catch it on the rebound with your scoop.
   c. First use one hand to catch the ball with the scoop and then try the other hand.

d. With a partner, throw the ball with the scoop in different overhand and underhand motions and catch the ball with the scoop.
e. Roll the ball to a partner with the scoop and try to catch the ball with the scoop.
f. Various modifications of volleyball, basketball, and softball which utilize the scoop can be developed.

BALLOONS

CONSTRUCTION:

Purchase large 5 cent circular balloons from a toy store, department store, or drug store. Use breadwrappers to fasten the balloons rather than tie them. Used in this manner, balloons can be deflated when finished and used repeatedly.

OBJECTIVES:

1. To develop eye-hand and eye-foot coordination.
2. To develop visual pursuit or tracking patterns.
3. To increase gross-motor coordination.
4. To develop concepts of laterality, directionality, position in space, and spatial relationships.

ACTIVITIES:

1. Singles
   a. Tap the balloon into the air with one hand. Then, use the other hand.
   b. Tap the balloon into the air with one of your fingers. Now try some of your other fingers.
   c. Tap the balloon into the air with both of your hands at the same time.
   d. Tap the balloon rapidly, then slowly. Does it take more or less force to tap the balloon rapidly? Slowly?
   e. Tap the balloon high into the air, then see how many times you can pass underneath the balloon before it touches the ground.
   f. Tap the balloon into the air with various body parts. Use your head, chin, shoulders, elbows, hips, feet, and knees. Use both sides of your body. See Figure 9.11.
   g. Tap the balloon into the air from various body positions. Lie down, sit down, kneel down, and stand up.
   h. Balance using one, two, three, or four bases of support while tapping the balloon into the air.
   i. Put a penny or other similar weight inside the balloon and blow it up. How does the added weight change the flight pattern of the balloon as you tap it?
   j. Mount the balloons on a pegboard and have the child pop them with darts. Establish safety guidelines with

FIGURE 9.11. Keep the balloon in the air by tapping it with various body parts.

the child before attempting these activities. Use smaller, less expensive balloons for this activity.

2. Partners.
   a. Volley the balloon with your partner using various body parts.
   b. Alternate right and left sides of your body as you pass the balloon to your partner.
   c. Use underhand and then overhand taps as you pass the balloon.
   d. Blow the balloon into the air to your partner.
   e. Tie the balloon on one of your ankles and walk, hop, or skip about the room, as you try to step on and break other children's balloons. Last one to keep his balloon from being broken is the winner of the game.

BUBBLES

CONSTRUCTION:

Soap bubbles can be purchased commercially or made by combining detergent and water in a container. Put the water in the container and add soap a little at a time until the right consistency is achieved. If bubble blowers are needed, purchase the plastic pop bottle stoppers that have the ring attachments. They come in packages of three or six and can be purchased at a grocery store. Coat hangers or thin pieces of wire can also be bent into the shape of a bubble blower. An inexpensive smoker's pipe can be converted into a suitable instrument for blowing soap bubbles if additional variations are desired.

OBJECTIVES:

1. To develop eye-hand coordination.
2. To help children develop better concepts of spatial awareness.
3. To develop body awareness of body image.
4. To increase a child's ocular control.

ACTIVITIES:

1. Blow some bubbles into the air. Have the child move after them and pop them with designated body parts. Have him use his index finger, thumb, ring finger, right hand, left hand, elbow, foot, and so on. See Figure 9.12.
2. Have the child blow some bubbles and then catch one back on the blower without popping the bubble.
3. Blow the bubbles in different ways. Blow air out of your mouth. Swing the bubble blower in the air in different ways.
4. Ask the child how many bubbles appear on each attempt. Try to set a record for the most bubbles on one try.
5. Have the child attempt to follow one bubble, until it lands or pops, while using only his eyes. Do now allow him to turn his head.
6. Vary the height and distance from which the bubbles are blown toward the child. Have the child follow the bubbles with his eyes as they move toward him.

FIGURE 9.12. Pop the soap bubbles by touching them with various body parts.

CANDLE AND SQUIRT GUN

CONSTRUCTION:

Obtain ½ to 1 inch buffet candles. Burned stubs, at least 2 to 4 inches tall, work fine, so children may bring these from home. Light a candle and drip wax onto a piece of cardboard to mount the candle on a base. Use small toy squirt guns purchased from a toy store to squirt out the candle flame. Perform this activity in an outdoor setting or in an open area where it will be easy to mop up the water when finished. See Figure 9.13.

FIGURE 9.13. Put out the candles with water from a squirt gun.

OBJECTIVES:

1. To develop eye-hand and fine-motor coordination.
2. To help a child learn to fixate on a target.

ACTIVITIES:

1. Light the candle. Have the child stand a few feet away from the candle. Ask him to extinguish the flame with the squirt gun. As the child's aim improves, increase the distance from the target.
2. Make targets on a chalkboard similar to

those on a dartboard. Have the child stand various distances away from the target and try to hit the "bull's eyes."

3. Try to squirt the water at a chalkboard and make designs, letters, or numbers.

TASK BOARDS

CONSTRUCTION:

Task boards are designed to help children learn to perform common, daily personal or household tasks with which they may be having trouble learning to perform. Buttons, snaps, zippers, shoe strings, hooks, and the like should be attached to material remnants and mounted on a piece of plywood, as seen in Figure 9.14. Window locks, door latches, chain locks, combination locks, door knobs, and the like can also be mounted on pieces of plywood and used for the same objectives.

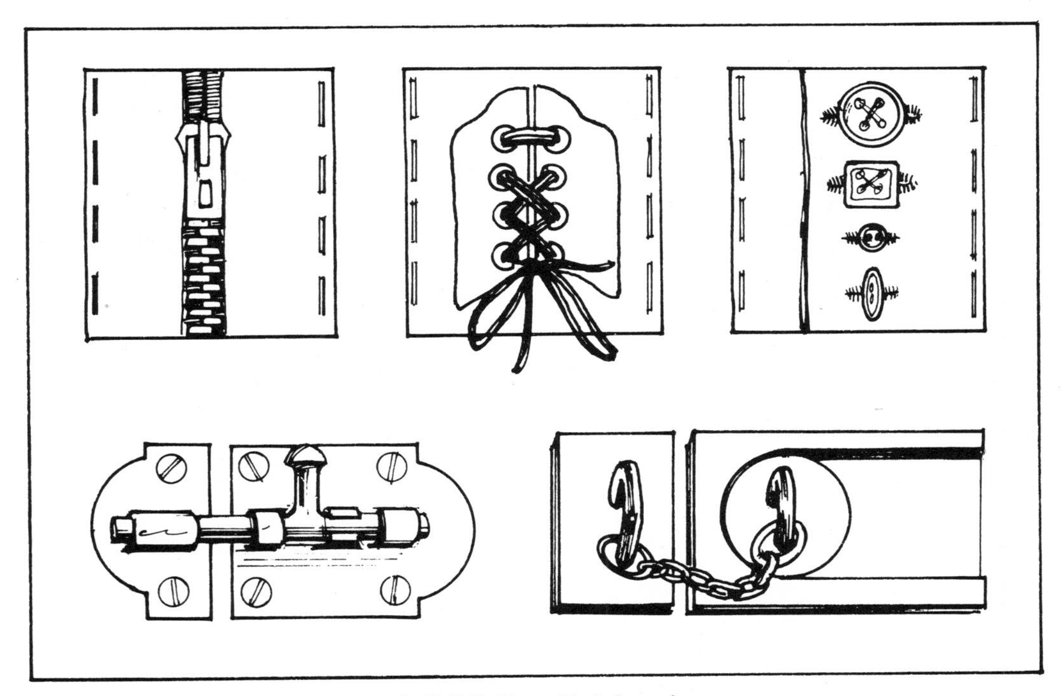

FIGURE 9.14. Task board.

OBJECTIVES:

1. To develop eye-hand and fine-motor coordination.
2. To learn to perform daily personal and household tasks.

ACTIVITIES:

1. Teach a child to tie his shoes, attach snaps, button buttons, and zip a zipper on the board. As he masters the board, see if he can perform the same tasks on his own clothes.
2. Give the child a board of hardware materials. Teach him to perform all of the tasks. Once he has mastered the board, take him around the house or school and have him attempt similar tasks.

CLOTHES PINS, MAGNETS, AND BOTTLE CAPS

CONSTRUCTION:

Collect wooden or plastic spring (clip) clothes pins and bottle caps from pop bottles and store them in a wide-mouth container, such as those that can be obtained from an ice cream store. Purchase some small, strong magnets from a hardware store. Fasten a string to the magnets and store them in the container

FIGURE 9.15. Use clothes pins, magnets, and bottle caps to develop eye-hand coordination.

also. One way to fasten the string to the magnets is to take a piece of nylon hose, wrap it around the magnet, and tie the string around the nylon hose to seal off the magnet. See Figure 9.15.

OBJECTIVES:

1. To develop eye-hand coordination.
2. To increase fine-motor coordination.
3. To increase a child's ability to follow a moving target.

ACTIVITIES:

1. Pour the contents of the jar onto the floor. Have the child use a clothes pin and place the caps one by one back into the jar.
2. Increase the difficulty level by asking the child to use his nondominant hand.
3. Have the child use both hands simultaneously by picking up a bottle cap with a clothes pin in each hand.
4. Perform the same exercises, using the magnet on a string.
5. Pick up a bottle cap with the magnet. Use the object as a target and perform all of the ocular pursuit activities related to the swinging ball found in another section of this book.
6. Use a container with a small mouth, such as a milk bottle or plastic bottle. Have the child drop bottle caps into the container from a kneeling or standing position.

NUTS, BOLTS, AND PIPES

CONSTRUCTION:

Obtain a variety of nuts and bolts from a hardware store. Construct a storage rack from a piece of 1" x 6" and use a piece of 2" x 4" for the base. Drill holes of various sizes in rows. Place nuts and bolts of the same size on each row. Graduate the rows from small to large. Obtain ½ or 1 inch pipe from whatever source you can. Discarded ends can often be obtained for free from plumbers or hardware stores. Have the pieces cut into 6-inch sections. Have both ends threaded. Purchase elbows of various types (90°, acute angle, obtuse angle, and "T") from a hardware store. Purchase a dead end plate for the pipe and attach it to a piece of plywood for a base. See Figures 9.16 and 9.17.

OBJECTIVES:

1. To develop eye-hand and fine-motor coordination.
2. To help children discriminate size.
3. To develop auditory and visual sequential memory.

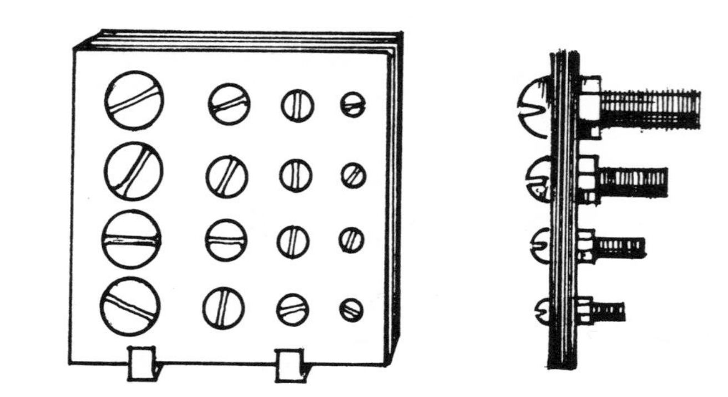

FIGURE 9.16. Make a nut and bolt board with graduated sizes.

ACTIVITIES:

1. Give the child the board of nuts and bolts and have him unscrew them from the board. Then, have him replace the nuts and bolts on the board in the same order.
2. Work with concepts of larger than and smaller than. For example, "take off the middle row of nuts and bolts. Then, find a row that is smaller than the middle row and take it off".
3. Take the nuts and bolts off the board or place them back as fast as you can. Time yourself and see if you can do it faster each time.
4. Using the dead end plate as a base, build a geometric tree with the pipes.
5. Use the pipes and elbows to create a design. Draw the design you have made on a piece of paper, then, draw it in the air.
6. Design a pattern with the pipes. Have the child create the same pattern. Remove it from his sight and have him create the same design from visual memory. Have the child close his eyes and create the pattern by using his tactile-kinesthetic sense only.
7. Place several pipes in a pile. Have the child put them together as quickly as he can. Time yourself to see if you can gradually assemble them faster.

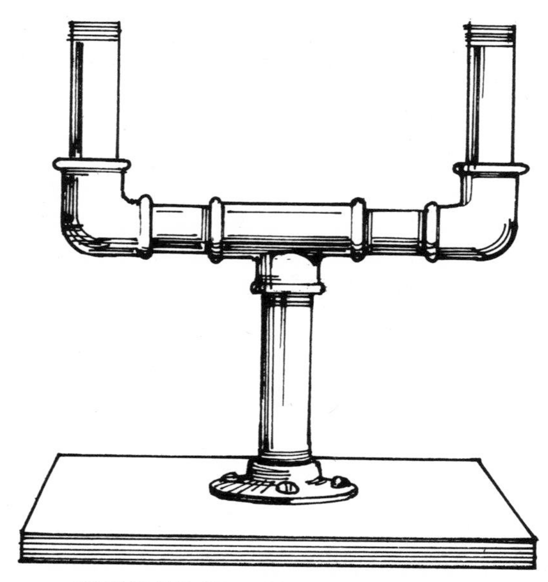

FIGURE 9.17. Make a pipe tree.

MARBLES AND WASHERS

CONSTRUCTION:

Obtain several dozen marbles and washers of various kinds and store them in a container, such as a discarded potato chip can.

OBJECTIVES:

1. To develop eye-hand and fine-motor coordination.
2. To teach children patterns of organization or classification.
3. To develop figure-ground perception.

ACTIVITIES:

1. Marbles
   a. Dump the marbles into a pile. Have the child categorize them by size, color, or composition.
   b. Dump the marbles into a pile. Have the child pick them up one at a time and put them back into the can as fast as he can. Use one hand, then the other hand. Use both hands, then alternate hands.
   c. Toss the marbles into cans, egg cartons, and so on, from various distances.
   d. Play a game of marbles with them.
   e. Use them with the marble track to develop a child's visual pursuit abilities (pp 84).
   f. Use them to play Chinese Checkers.
   g. Put several into a can to make a rhythmical instrument.
2. Washers
   a. Dump the washers into a pile. Have the child sort them into categories, such as all rubber washers, metal washers, or washers with the same size hole.
   b. Dump the washers into a pile. Have the child pick them up, one at a time, and put them back into the can as fast as he can. Use one hand, then the other. Use both hands, then alternate hands.
   c. Toss the washers into cans, egg cartons, and the like from various distances.
   d. String the washers on coat hangers to make a tambourine for various rhythmical activities.
   e. Play checkers, tic-tac-toe, finger shuffle board, caroms, or tiddley-winks with them.

CUTTING AND PASTING

CONSTRUCTION:

Various kinds of paper, paste, and scissors are needed for this activity. The construction is the child's and/or teacher's responsibility.

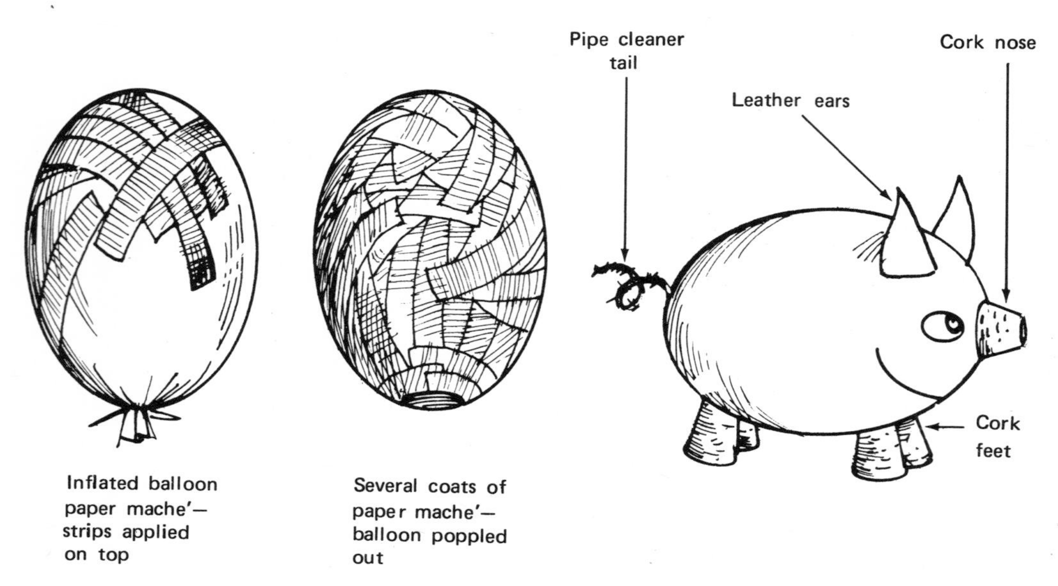

FIGURE 9.18. Make a paper maché animal.

OBJECTIVES:

1. To develop eye-hand coordination.
2. To increase fine-motor control.
3. To develop ocular fixations and visual pursuit.

ACTIVITIES:

1. Have the child make a poster with every geometric form he knows. After he has cut out each form, have him paste them on a large piece of background paper and label each form, if desired. Various types of designs can be made by pasting the forms together in different orientations.
2. Give the child some flour and water to use to make his own paste. Then, have him cut newspaper into 2 inch strips with a scissors. Have the child run the paper strips through the glue and paste them on a blown up balloon that is coated with glue. Let it dry and then place two or three mory layers of glue on the balloon. Stick a pick through the paper mold and pop the balloon. Now, with a little creativity the child can complete a paper mache sculpture. For example, paint the body and add ears, legs, and a face to make a bank. See Figure 9.18.
3. Give the child a coloring book page. Give him a pair of scissors and have him cut out the main figure in the picture. As he improves, have him cut out comic book characters or newspaper pictures.
4. If the child needs to develop strength in his preferred hand, use thicker materials. Magazine covers, construction paper, or thin cardboard are good materials for this exercise.
5. Any arts and crafts project, where the child is required to work with his hands, will develop eye-hand and fine-motor coordination. Paper tolling, quillery, decoupage, macrame, and making models all

offer a challenge as the child increases the quality of his performance. Craft stores provide the materials and directions, if needed.

PAPER AND PENCIL EXERCISES

CONSTRUCTION:

A marking pencil, transparency, notebook, and some paper are needed. Make up several workbook pages with tasks that are designed to increase the child's fine-motor coordination (writing behavior). Use a piece of clear acetate or plastic to cover the workbook pages to make them durable. Keep the pages in a notebook.

OBJECTIVES:

1. To develop eye-hand and fine-motor coordination.
2. To develop manual encoding abilities.
3. To develop form perception.

ACTIVITIES:

1. Make up some workbook pages that require the child to:
   a. Draw between the lines in a horizontal, vertical, diagonal, and curved direction.
   b. Draw on the line in a horizontal, vertical, diagonal, and curved direction.
   c. Complete the missing parts of the drawings. Use incomplete geometric forms, human faces, human bodies, houses, and the like.
   d. Complete the numbered dots to make a total picture. See Figure-9.19.
2. Have the child play tic-tac-toe with a friend while using a piece of paper and a pencil.
3. Give the child a piece of paper. The size may vary from an 8½ x 11" piece of notebook paper to a sheet of newspaper. Using only one hand, have the child crumple or wad the paper into a ball. This will develop the hand muscles. After the paper is shaped into a ball, use masking tape to bind it. The ball can now be used for throwing and catching activities.

** For additional suggestions on writing activities, refer to Chalkboard and Template Activities in another section of this book (pp 68-69 and 140).

STRAWS AND TOOTHPICKS

CONSTRUCTION:

Purchase a box of drinking straws and toothpicks from a grocery store. Store them in a container to prevent them from getting scattered and lost.

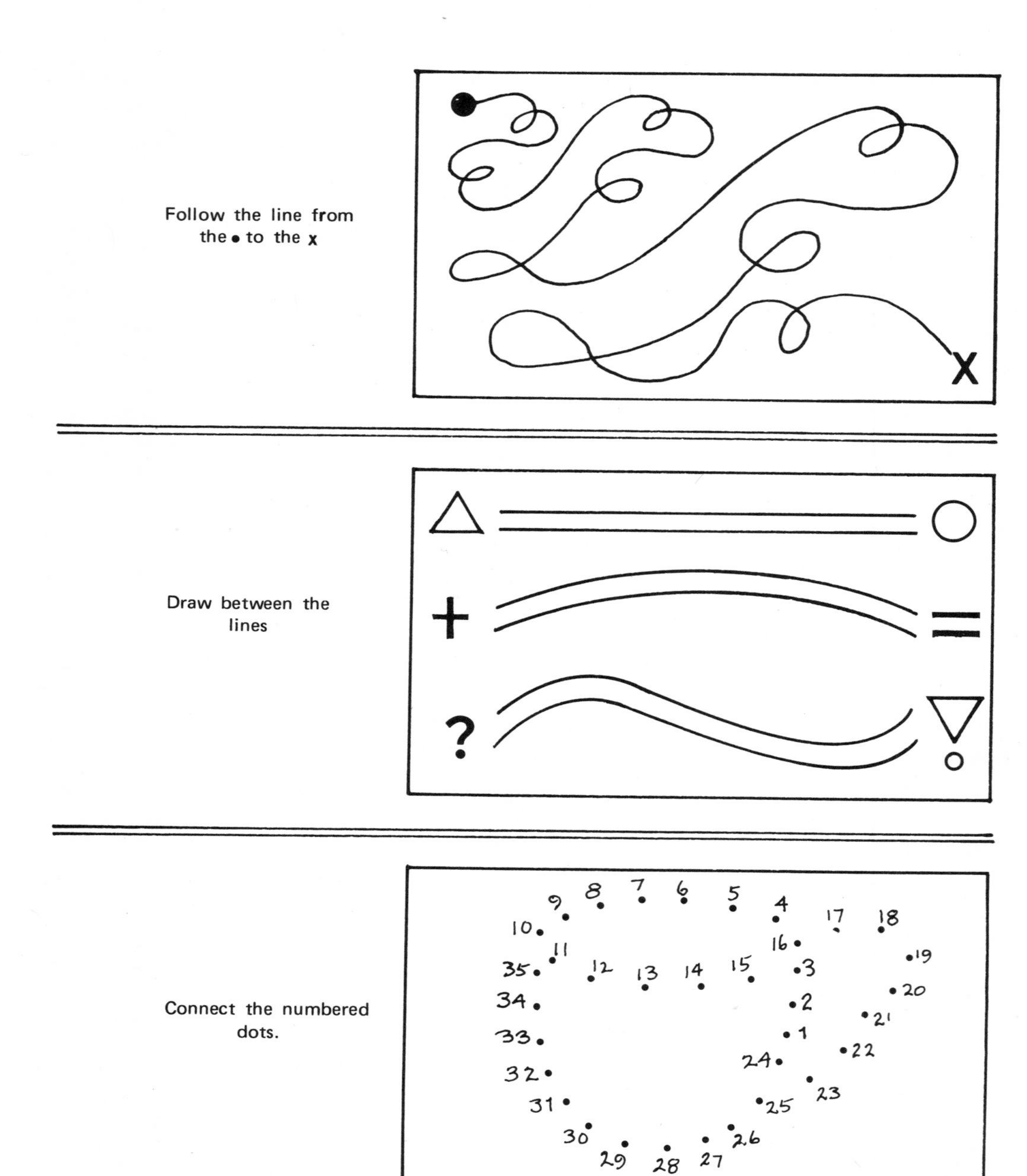

FIGURE 9.19. Examples of paper and pencil exercises.

OBJECTIVES:

1. To develop eye-hand and fine-motor coordination.
2. To teach children better concepts of position in space and spatial relationships.

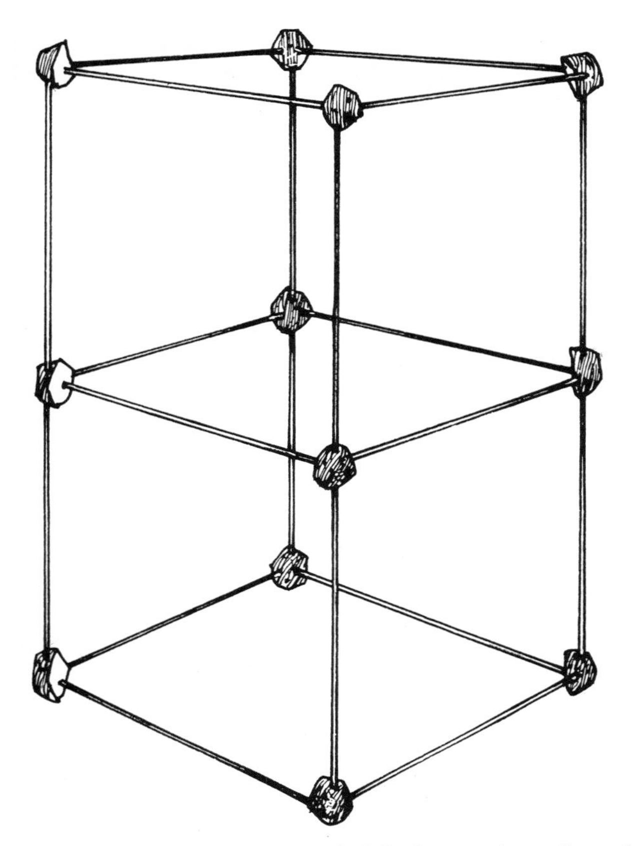

FIGURE 9.20. **Potato pieces and toothpicks form a three dimensional tower.**

3. To help children learn to auditorily and visually decode messages.
4. To develop form perception.
5. To develop manual encoding abilities.

ACTIVITIES:

1. Dump the toothpicks or straws into a pile. Pick them up, one at a time, and put them back into the container. Use right hand only, left hand only, both hands at the same time, and alternate hands. Keep time and see if you can gradually increase your speed.
2. Play a game of pick-up-sticks with the toothpicks or straws. After dumping them into a pile, try to pick them up, one at a time, without moving any of the others.
3. Make up geometric designs with the straws or toothpicks by laying them in various directions. Make a cross, square, triangle, diamond, and other more complex shapes. Design houses and other forms with the straws and toothpicks.
4. Form the straws or toothpicks into numbers or letters of the alphabet. Make up number or letter series upon the teacher's instruction.
5. Repeat steps two and three, but this time change the directions. Give auditory directions or a pictorial design and have the

child make the design with the straws or toothpicks by relying on his auditory or visual memory. Have the child close his eyes and make the design using his tactile-kinesthetic sense.

6. Cut a potato into small cubes or pull clay into small pieces. Stick the straws or toothpicks into the potato or clay pieces and make a three-dimensional design. Make it as intricate as you like. Glue can be used to hold the straws or toothpicks together for a permanent design. Popcycle sticks may be used instead of the straws or toothpicks. See Figure 9.20.

FOOT LAUNCHERS

CONSTRUCTION:

For each foot launcher, purchase a piece of 1 x 6 lumber, 3 feet long. Use some scrapwood to create a 1 x 6" fulcrum, two inches tall. Nail the fulcrum 6 inches from one end of the 3 foot board, as seen in Figure 9.20. The board can be painted with a foot at the short end and a beanbag at the long end. This gives the child a visual clue.

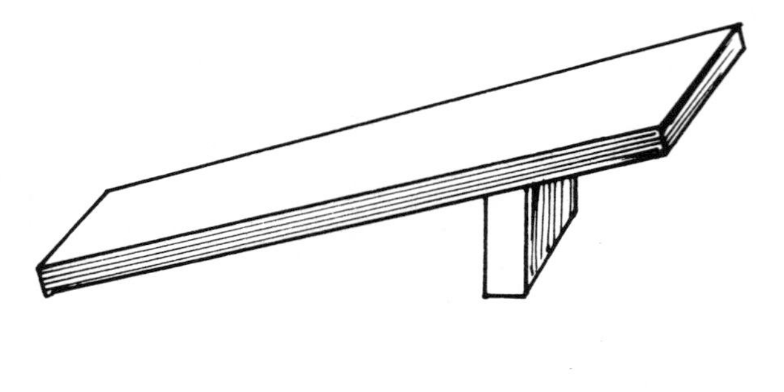

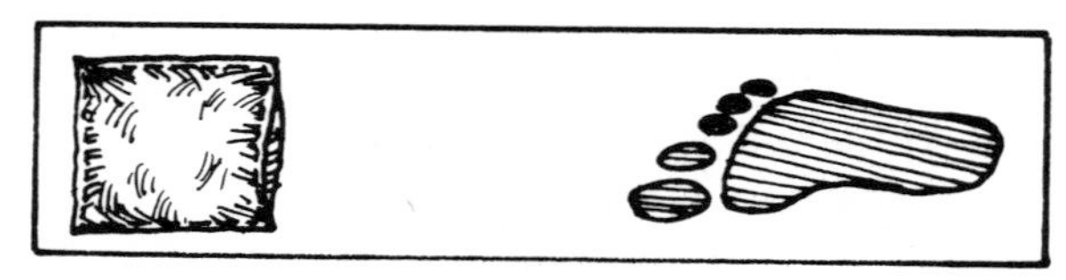

FIGURE 9.21. Foot launcher.

OBJECTIVES:

1. To develop eye-hand, eye-foot, and gross-motor coordination.
2. To increase a child's self-concept and spatial awareness.

ACTIVITIES:

1. Place a beanbag on the end of the board. Have the child stamp on the raised end to launch the beanbag. Then, have the child try to catch the beanbag while it is in the air.
2. Vary a launched object, such as a yarn ball, fleece ball, or paper ball.
3. Have the child attempt to launch the

object to specific heights, such as his waist, chest, chin, eyes, and so on.

4. Have the child attempt to catch the object with different body parts, such as his palm, back of hand, wrist, elbow, or foot.

Appendix A

1. **AAHPER,** Perceptual-Motor Foundations: A Multidisciplinary Concern. Washington, D.C., 1968.
2. **AAHPER,** Foundations and Practices in Perceptual-Motor Learning—A Quest for Understanding. Washington, D.C., 1971.
3. **AAHPER,** Annotated Bibliography on Perceptual-Motor Development. Washington, D.C., 1973.
4. **Ayres, Jean,** Sensory Integration and Learning Disorders. Los Angeles, California: Western Psychological Services, 1972.
5. **Barsch, Ray,** A Movigenic Curriculum. Madison, Wisconsin: Bureau of Audiovisual Instruction, University of Wisconsin, 1965.
6. **Bush, Wilma J., and Giles, Marian T.,** Aids to Psycholinguistic Teaching. Columbus, Ohio: C. E. Merrill, 1969.
7. **Chaney, Clara, and Kephart, N. C.,** Motoric Aids to Perceptual Training. Columbus, Ohio: C. E. Merrill, 1968.
8. **Cratty, Bryant J.,** Developmental Sequences of Perceptual-Motor Tasks. Freeport, Long Island, New York: Educational Activities, Inc., 1967.
9. **Cratty, Bryant J.,** Intelligence in Action. Englewood Cliffs, New Jersey: Prentice-Hall, 1970.

10. **Cratty, Bryant J.,** Perceptual-Motor Behavior and Educational Processes. Springfield, Illinois: Charles C Thomas, 1970.
11. **Cratty, Bryant J., and Martin, Sister M. M.,** Perceptual-Motor Efficiency in Children. Philadelphia: Lea and Febiger, 1969.
12. **Cratty, Bryant J.,** Perceptual and Motor Development in Infants and Children. New York: Macmillan, 1970.
13. **Cruickshank, William M., et al.,** A Teaching Method for Brain Injured and Hyperactive Children. New York: Syracuse University Press, 1961.
14. **Delacato, Carl,** The Diagnosis and Treatment of Speech and Reading Problems. Springfield, Illinois: Charles C Thomas, 1963.
15. **Early, George H., "Low Level Functional Deficits in Learning Disabled Children,"** Academic Therapy 8:231-34, Winter, 1972.
16. **Early, George H.,** Perceptual Training in the Classroom. Columbus, Ohio: C. E. Merrill, 1969.
17. **Estes, R. E., and Huizinga, R. J., "A Comparison of Visual and Auditory Presentations of a Paired-Associate Learning Task with Learning Disabled Children,"** Journal of Learning Disabilities 7:44-51, January, 1974.
18. **Frierson, Edward C., and Barbe, Walter B.,** Educating Children with Learning Disabilities. New York: Appleton-Century-Crofts, 1967.
19. **Frostig, Marianne, and Horne, D.,** The Frostig Program for the Development of Visual Perception. Chicago: Follett Educational Corporation, 1964.
20. **Furth, Hans G., and Wachs, Harry,** Thinking Goes to School. New York: Oxford University Press, 1974.
21. **Getman, G. N., et al.,** Developing Learning Readiness. St. Louis: McGraw-Hill, 1968.
22. **Godfrey, Barbara, and Kephart, N. C.,** Movement Patterns and Motor Education. New York: Appleton-Century-Crofts, 1969.
23. **Hammill, Donald D., and Bartel, Nettie R.,** Educational Perspectives in Learning Disabilities. New York: Wiley, 1971.
24. **Harvat, Robert W.,** Physical Education for Children with Perceptual-Motor Learning Disabilities. Columbus, Ohio: C. E. Merrill, 1971.
25. **Heath, Edward, and Early, George H., "Intramodal and Intermodal Functioning of Normal and Learning Disabled Children,"** Academic Therapy 9:133-49, Winter, 1973.
26. **Karnes, Merle B.,** Helping Young Children Develop Language Skills. Arlington, Virginia: The Council for Exceptional Children, 1968.
27. **Kephart, Newell C.,** The Slow Learner in the Classroom. Columbus, Ohio: C. E. Merrill, 1971.
28. **Lerch, Harold A., et al.,** Perceptual-Motor Learning-Theory and Practice. Palo Alto, California: Peek Publications, 1974.
29. **Lerner, Janet W.,** Children with Learning Disabilities. Boston: Houghton-Mifflin, 1971.
30. **Levy, Harold B.,** Square Pegs, Round Holes. Boston: Little, Brown and Company, 1973.
31. **McCarthy, John, and McCarthy, J. F.,** Learning Disabilities. Boston: Allyn and Bacon, 1970.
32. **McCulloch, Lovell,** Handbook for Developing Perceptual-Motor and Sensory Skills. Austin, Texas: Austin Writer's Group, 1975.
33. **Mourozis, A., et al.,** Body Management Activities: A Guide to Perceptual-Motor Training. Cedar Rapids, Iowa: Nissen Company, 1970.
34. **Myers, P. I., and Hammill, D. D.,** Methods for Learning Disorders. New York: Wiley, 1969.
35. **Radler, D. H., and Kephart, N. C.,** Success Through Play. New York: Harper and Row, 1960.

36. **Sage, George, H.,** Introduction to Motor Behavior: A Neuropsychological Approach. Reading, Massachusetts: Addison Wesley, 1971.
37. **Saphier, J. D., "The Relation of Perceptual-Motor Skills to Learning and School Success,"** Journal of Learning Disabilities, 6:56–65, November, 1973.
38. **Simpson, Dorothy, M.,** Learning To Learn. Columbus, Ohio: C. E. Merrill, 1968.
39. **Valett, Robert,** Programming Learning Disabilities. Palo Alto, California: Fearon Publishers, 1969.
40. **Waugh, Kenneth W., and Bush, Wilma J.,** Diagnosing Learning Disorders. Columbus, Ohio: C. E. Merrill, 1971.
41. **Williams, H. G., "Perceptual-Motor Development in Children,"** A Textbook of Motor Development. Dubuque, Iowa: Wm. C. Brown Company, Charles B. Corbin, Editor, 1973.
42. **Wunderlich, Ray C.,** Kids, Brains and Learning. St. Petersburg, Florida: Johnny Reads, Inc., 1970.
43. **Wunderlich, Ray C.,** Allergies, Brains, and Children Coping. St. Petersburg, Florida: Johnny Reads, Inc., 1973.

Appendix B

And So They Move. *(16mm, b&w, 20 min.)* Available Audio-Visual Center, Michigan State University, East Lansing, Michigan 48824.

Many practical and meaningful activities on fundamental movement experiences for physically handicapped children are presented with accompanying narration of the theoretical value of the activities. Suggestions are included for sequence in programming based on a problem solving approach.

Anyone Can: Learning Through Motor Development. *(16mm, color, sd., 27 min.)* Available Bradley Wright Films, 309 North Duane Avenue, San Gabriel, California 91775.

Four short films on one reel designed to help teachers develop a balanced program of motor activities for atypical children. Activities included are: rope skills, ball handling, the stegel, and the trampoline.

A Time To Move. *(16mm, b&w, sd., 30 min.)* Available Early Childhood Productions, Box 352, Chatsworth, California 91311.

A film focused on the meaning of movement for the 3 and 4 year old. Movement is the first and deepest language of the child for its own sake and for what it achieves.

Blocks–A Medium for Perceptual Learning., *(16mm, color, sd., 18 min.)* Available Campus Film Productions, 20 East 46th Street, New York, N.Y. 10017.

Blocks have long been recognized as an essential medium for learning. They foster physical, social, and emotional growth. Block building also provides a framework for many academic situations. However, basic to these are a child's perceptual learnings, derived from how he perceives the blocks from which he works and the space in which he builds.

Bridges To Learning. *(16mm, color, sd., 30 min.)* Available Palmer Films Inc., 611 Howard Street, San Francisco, California.

Illustrates the organization and administration of a K-6 physical education program, with emphasis on perceptual training and innovative curriculum related to skills, games, and sports, including evaluation techniques.

Bright Boy, Bad Scholar. *(16mm, b&w., sd., 25 min.)* Available Indiana University Film Library, Bloomington, Indiana 47401.

The film makes an interview approach to understanding learning disabilities and presents many case studies as examples of children who possess symptoms of perceptual-motor problems.

Creative Body Movements. *(16mm, color, sd., 11 min.)* Available Martin Moyer Productions, 900 Federal Avenue, East Seattle, Washington, 98102.

Shows how children in the primary grades can express themselves through movement, using a perceptual-motor and problem solving approach.

Developmental Physical Education. *(16mm, color, sd., 28 min.)* Available Simensen and Johnson, Education Consultants, Box 34, College Park, Maryland 20740.

A film developed by Louis Bowers, University of South Florida, depicting the development of balance, laterality, directionality, body image, spatial awareness, and visual perception among mentally retarded children.

Dyslexia: Prevention and Remediation-A Classroom Approach. *(16mm, color, sd., 21 min.)* Available The Audiovisual Center, Prince George's County Public Schools, 4800 Varnum Street, Beadensbury, Maryland 20710.

The film shows developmental training techniques stressing both how dyslexic children can learn and how future teachers of those now working with these children can use these techniques to teach them.

Early Recognition of Learning Disabilities. *(16mm, color, sd., 30 min.)* Available Churchwell Films, Department of HEW, Station K, Atlanta, Georgia 30324.

The film emphasizes the importance of early recognition and programming to help those children who have perceptual-motor problems.

In, Out, Up, Down, Over, Upside Down. *(16mm, color, sd., 8 min.)* Available ACI Films, Inc., 35 West 45th Street, New York, N.Y. 10036.

A film that deals with the development of the concept of directions and body positions in space through movement activities of children.

Learning Disabilities. *(16mm, b&w, sd., 30 min.)* Available NBC Educational Enterprises, 30 Rockefeller Plaza, New York, N.Y.

The film depicts many examples of symptoms of learning disabilities, explores the problems arising from these, and illustrates specific exercises used to remedy deficiencies.

Learning To Learn. *(16mm, color, sd., 24 min.)* Available Wardell Associates, 49 Pinckney Street, Boston, Massachusetts.

The film identifies children who are having trouble learning in a normal school situation as unforthcoming and inconsequential. The film presents a behavior modification approach through the Flying Start To Learning Program to help these children overcome their problems.

Movigenic Curriculum. *(16mm, b&w, sd., 30 min.)* Available Bureau of Audio Visual Instruction, University of Wisconsin, 1327 University Avenue, Madison, Wisconsin 53706.

The film illustrates the twelve dimensions or constructs of Ray Barsch's movigenic curriculum to help children with perceptual-motor problems.

Physical Education-Lever To Learning. *(16mm, color, sd., 24 min.)* Available Stuart Finley Films, 3428 Mansfield Road, Falls Church, Virginia 22041.

The film shows educable mentally retarded boys and girls from a public school special education program taking part in a vigorous and varied program emphasizing the development of motor skills and physical fitness through the use of innovative equipment.

Sensorimotor Training. *(16mm, color, sd., 24 min.)* Available Valdhere Films, 3060 Valleywood Drive, Kettering, Ohio.

Describes philosophy and training methods for helping preschool children develop sensory skills and physical coordination in the Dayton Public Schools, Dayton, Ohio.

Thinking, Moving, Learning. *(16mm, color, sd., 30 min.)* Available Bradley Wright Films, 309 North Duane Avenue, San Gabriel, California 91775.

A comprehensive movement program at the preschool and early elementary levels can help children develop a positive self concept, spatial awareness, visual perception, and other concepts that help them achieve a readiness level for success in school.

Visual Perception and a Failure to Learn. *(16mm, b&w, sd., 30 min.)* Available Indiana University Film Library, Bloomington, Indiana 47401.

The film focuses on the Frostig Developmental Test of Visual Perception. Eye-hand coordination, figure-ground perception, perceptual constancy, position in space, and spatial relationships are discussed.

Why Billie Couldn't Learn. *(16mm, color, sd., 40 min.)* Available California Association for Neurologically Handicapped Children, Film Director, Box 604, Main Office, Los Angeles, California 90053.

Focuses on the diagonsis and teaching techniques used in a special classroom for neurologically handicapped children.

Appendix C

TESTS

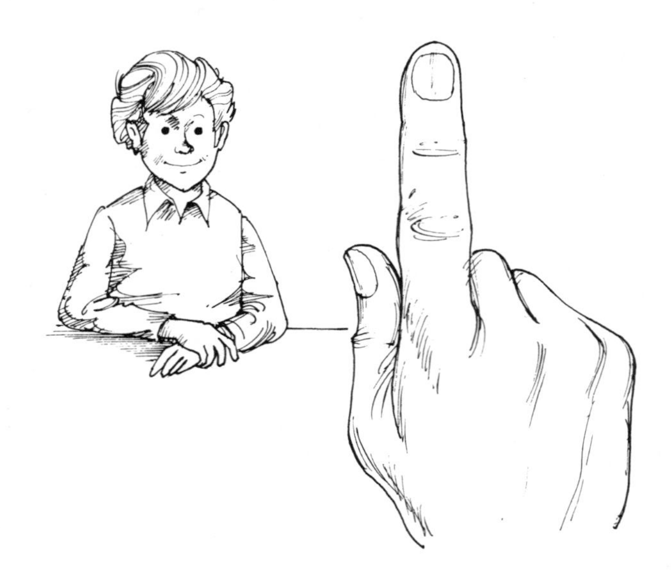

Ayres, Jean, Southern California Sensory Integration Tests, Western Psychological Services, Los Angeles, California.

Beery, K. E., Developmental Test of Visual-Motor Integration, Follett Educational Corporation, Chicago, Illinois.

Bender, Laura, Bender Visual Motor Gestalt Test, The Psychological Corporation, New York, N.Y.

Brenner, Anton, Anton Brenner Developmental Gestalt Test of School Readiness, Western Psychological Services, Los Angeles, California.

Cheves, R., Pupil Record of Educational Behavior, Teaching Resources Corporation, Boston, Massachusetts.

Doll, E., Lincoln Oseretsky Motor Proficiency Tests, American Guidance Service, Inc., Circle Pines, Minnesota.

Foster, C. R., et. al., Assessment of Children's Language Comprehension, Consulting Psychologists Press, Palo Alto, California.

Frostig, Marianne, The Frostig Developmental Test of Visual Perception, Consulting Psychologists Press, Palo Alto, California.

Gates, Arthur, and MacGinitie, Walter H., Gates-MacGinitie Reading Tests, Western Psychological Services, Los Angeles, California.

Goodenough, Florence, et. al., Minnesota Preschool Scale, Western Psychological Services, Los Angeles, California.

Ilg, Frances L., and Ames, Louise Bates, School Readiness: Behavior Tests Used at Gessell Institute, Western Psychological Services, Los Angeles, California.

Jastak, J. F., and Jastak, S. R., Wide Range Achievement Test, Revised, Western Psychological Services, Los Angeles, California.

Katz, J., Kindergarten Auditory Screening Test, Follett Educational Corporation, Chicago, Illinois.

Kirk, S., et. al., Illinois Test of Psycholinguistic Abilities, Revised, Western Psychological Services, Los Angeles, California.

Minnesota Percepto-Diagnostic Test, Western Psychological Services, Los Angeles, California.

Peabody Picture Vocabulary Test, American Guidance Service Publishers, Circle Pines, Minnesota.

Perceptual Testing and Training Kit for Kindergarten and First Grade Teachers, Winter Haven Lions Research Foundation, Inc., Winter Haven, Florida.

Point, Arthur, Arthur Point Scale of Performance Tests, J. A. Preston Corporation, New York, N.Y.

Roach, E., and Kephart, N., The Purdue Perceptual-Motor Survey, Charles E. Merrill, Columbus, Ohio.

Semel, E., Sound, Order, Sense, Follett Educational Corporation, Chicago, Illinois.

Slosson, Richard, Slosson Intelligence Test, Western Psychological Services, Los Angeles, California.

Stanford-Binet Intelligence Scale, Combined L-M Form, Houghton Mifflin, Boston, Massachusetts.

Valett, Robert, Valett Developmental Survey of Basic Learning Abilities, Fearon Publishers, Palo Alto, California.

Wepman, J., Auditory Discrimination, Memory Span, and Sequential Memory Tests, Western Psyhological Services, Los Angeles, California.

Wechsler Intelligence Scale for Children, Psychological Corporation, New York, N.Y.

Appendix D

Manufacturer	*Address*	*Specialties*
Bernell Corporation	422 East Monroe St. South Bend, Indiana 46601	Vision testing and training equipment
Creative Playthings	Princeton, New Jersey 08540	Developmental equipment for young children
Occupational Work Experience Program	348 West First Street Dayton, Ohio 45402	Developmental equipment for young children
Developmental Learning Materials	7440 Natchez Avenue Niles, Illinois 60648	Developmental equipment for young children
Follett Publishing Company	1010 West Washington Blvd. Chicago, Illinois 60607	Early childhood education: materials for assessment and instruction

John R. Green Company	411 West Sixth Street Covington, Kentucky 41011	School supply company with a good assortment of equipment for early childhood
Ideal School Supply Company	11000 S. Lavergne Avenue Oak Lawn, Illinois 60453	Equipment for special education and perceptual-motor development
Lafayette Instrument Company	P.O. Box 1279 52 By-Pass Lafayette, Indiana 47902	Sensory and perceptual-motor testing and training equipment including tachistoscopes
Nissen Gymnastics Equipment Company	930 27th Avenue Cedar Rapids, Iowa 52406	Educator line of gymnastics equipment-low balance beam, trampoline
J. A. Preston Corporation	71 Fifth Avenue New York, New York 10003	Materials for exceptional children and youth: perceptual-motor and developmental physical education equipment
Skill Development Equipment Company	1340 N. Jefferson Anaheim, California 92806	Gross motor developmental equipment for early childhood and special education

Index